A DETAILED HISTORY OF
RAF MANSTON
1941-1945

A DETAILED HISTORY OF RAF MANSTON 1941-1945

Invicta—The Undefeated

JOE BAMFORD AND JOHN WILLIAMS
WITH PETER GALLAGHER

This book is dedicated to my late partner, Janice Saunders; without her help, this book would not have been possible. Also dedicated to the people of many different political persuasions who have campaigned to re-open Manston airport.

Fonthill Media Language Policy

Fonthill Media publishes in the international English language market. One language edition is published worldwide. As there are minor differences in spelling and presentation, especially with regard to American English and British English, a policy is necessary to define which form of English to use. The Fonthill Policy is to use the form of English native to the authors. Joe Bamford, John Williams, and Peter Gallagher were born and educated in the United Kingdom; therefore British English has been adopted in this publication.

Fonthill Media Limited
Fonthill Media LLC
www.fonthillmedia.com
office@fonthillmedia.com

First published in the United Kingdom and the United States of America 2016

British Library Cataloguing in Publication Data:
A catalogue record for this book is available from the British Library

Copyright © Joe Bamford, John Williams, and Peter Gallagher 2016

ISBN 978-1-78155-096-0

The right of Joe Bamford, John Williams, and Peter Gallagher to be identified as the authors of this work has been asserted by them in accordance with the Copyright, Designs and Patents Act 1988.

All rights reserved. No part of this publication may be reproduced, stored in a retrieval system or transmitted in any form or by any means, electronic, mechanical, photocopying, recording or otherwise, without prior permission in writing from Fonthill Media Limited

Typeset in 10pt on 13pt Sabon
Printed and bound by CPI Group (UK) Ltd, Croydon, CR0 4YY

Acknowledgements

A great deal of thanks goes to our researcher Peter Gallagher for searching the Operational Record Books of Manston at the National Archives. Without Peter's help, we would have spent a great deal of time and money travelling to London; instead, we have been able to devote more time to the process of writing the book.

Inspiration for this and our other two books in the trilogy about the history of RAF Manston has been provided by a number of people, but special thanks go to Wing Commander Chris Thorpe, former Warrant Officer Derek Crow-Brown, Marcus Russell, and Peter Turner. We also thank Dougie Cockerell at the RAF Manston History Museum and Matt from the Spitfire Memorial Museum. The fact that Manston has two museums that are very popular with tourists goes to prove that there is a great deal of interest in this historical airfield that dates back to May 1916.

There are a number of publications that we must acknowledge; amongst them are Rocky Stockman's *History of RAF Manston* and *Flying Through Fire* by Geoffrey Williams. Also *Glider Pilots at Arnhem* by Mike Peters and Luuk Buist provided useful information about the ill-fated Operation Market Garden in September 1944.

One mystery still remains and that concerns the construction of the control tower. There are a number of documents concerning the rebuilding of the airfield in 1942–43 leading to the opening of the new 9,000-foot runway in April 1943, but there is no mention of the control tower. Despite making enquiries with the Air Historical Branch and the RAF Museum, no new information has been found and the authors would welcome any details that readers might be aware of.

On a final note, as this publication goes to press, the future of Manston Airport still hangs in the balance. Having failed to agree to RiverOak becoming Thanet District Council's indemnity partner for a Compulsory Purchase Order, the company announced that it is to go alone and take out a Development Consent Order. We wish them all the best and look forward to the reopening of Manston Airport in the near future.

CONTENTS

Acknowledgements   5
Introduction   9

1	Manston on the Offensive	11
2	The Flying Dutchmen	17
3	Channel Stop	21
4	The Lone Wolf	31
5	Battling Beauforts	35
6	Operation Fuller	38
7	Calm after the Storm	46
8	The Dieppe Raid	54
9	Any Port in a Storm	59
10	The Fortunes of War	68
11	A Distinguished Visitor	75
12	Focke-Wulf Galore	82
13	Emergency Runway	87
14	Modern Manston	92
15	The Fiftieth Month of the War	101
16	The Year of Victory?	111
17	New Faces, New Runway	120
18	A Bad Day for Thanet	129
19	Towards D-Day	133
20	A Safe Haven	144
21	Squadron Leader Joseph Berry: The V-1 Champion	148
22	The Deadly Menace	152
23	Remembering Arnhem	158
24	Towards the End	171
25	The Year of Victory	176
26	The Final Chapter	184

Introduction

The main difference between this book and the first two volumes is that this book lacks the detail that they contained because of constraints on time and space. This lack of detail is mainly due to the high volume of events that took place during the Second World War, which far exceed the number that took place in the years covered by the previous volumes.

A huge number of fighter squadrons were based at Manston during the period 1941 to 1945, and while many used it as a forward base, arriving in the morning and departing for their own airfields at dusk, others were long-term residents. There were so many accidents and incidents that to mention them all would require a book on the scale of the *Encyclopaedia Britannica*. In the 1930s, a single day's movements and events at Manston was often covered by just a few lines in the Operational Record Book (ORB), but by 1941 and the consecutive war years a single day's events covered three or four pages. What we have tried to do is to filter the records and choose the most interesting events, particularly those that took place at RAF Manston.

Most of the action in this account directly concerns movements, events, and incidents that occurred at RAF Manston, but it also covers some that happened off the airfield. RAF Manston is located on the Isle of Thanet, bounded by the Wantsum Channel that ran from Richborough and Reculver. The three major towns are Ramsgate, Margate, and Broadstairs, which always had close connections to the RAF station; where the authors have thought it necessary, we have included accounts of events in those towns that concern the RAF and the war.

Some of the topics covered in our account, such as the Channel Dash, the Dieppe Raid, and 617 Squadron, have been the subject of books in their own right; in this account, we are only able to give just a basic outline of such actions. As a result of the extensive entries, this edition should be considered as just a history of RAF Manston, rather than a much more detailed account, as appeared in the previous two books. Nevertheless, it is an accurate and interesting account of one of the most important periods in the station's existence.

1

Manston on the Offensive

The year 1941 began in much the same way as 1940 had ended, with further attacks on the airfield by the Luftwaffe. During the afternoon of 11 January, two Bf 109s suddenly appeared out of the clouds. There might have been another couple of enemy fighters hidden by cloud, but the two that appeared each dropped a 250-kg bomb. One of these bombs landed on the north-east corner of the airfield, narrowly missing some Spitfires, while the second one landed in the centre of the airfield.

The two Bf 109s had appeared flying in line astern formation, and as soon as they dropped their bombs they climbed straight back up into the cloud and disappeared. The AA defences had gone into action and it was claimed that one of the enemy aircraft had been hit by fire from a Lewis gun. One bomb had been dropped from such a low height that witnesses claimed it had bounced along the ground for several seconds before finally exploding. Fortunately, the only damage was to an unserviceable Spitfire that was in the hangar, and it had been hit by shrapnel that had been blasted through a tiny gap in the hangar doors.

On 9 January, 92 Squadron arrived at Manston from Biggin Hill and it was the first resident unit to be posted to the station since the Battle of Britain the previous year. The squadron was commanded by a Canadian, Squadron Leader Johnny A. Kent, AFC, and was still equipped with the Spitfire Mk I, which it had been flying since March 1940. However, within a short while, it would become the first squadron to be equipped with the Spitfire Mk VB.

Squadron Leader Kent did not like 92 Squadron's way of doing things; soon after taking over, he called all the squadron's pilots together and gave them a 'pep' talk. He accused them of being an ill-controlled rabble, but said that he was going to change things—he particularly criticised Flight Commander Brian Kingscombe, who was accused of breaching the King's regulations as he had recently flown on a sortie while still being medically unfit. Most of the pilots were upset by Kent's remarks because the Squadron was one of the highest-scoring units and its discipline in the air was among the best in Fighter Command. It was not a good start for Kent's command, but 92 Squadron's pilots would not be brow-beaten and it was he who eventually had to fit into the squadron's routines.

On 28 January, a single Ju 88 attacked the dispersal headquarters that had been moved off the airfield into buildings in the town of Westgate, and once again the Luftwaffe used a low cloud base for cover. The Ju 88 dropped three bombs before the sweeping back up into the cloud—the only damage was to two brick buildings that had been used as lodges.

No. 92 Squadron flew four patrols on 2 February, but heavy snow had fallen and the pilots were not very comfortable at Manston because a small number of Blenheim crews from 59 Squadron had moved in and things were rather crowded. The following morning after breakfast, the Squadron sent up four Spitfires to search for enemy aircraft that had bombed Ramsgate, but flew uneventful patrols. That afternoon, however, Fg Off. Tony Bartley shot down a He 111 into the sea near Southend and it was the Squadron's first victory since December 1940. It was claimed that the cannon that had been recently fitted to the Spitfires had been a great success.

At 11.45 a.m. on 7 February, news was received that an aircraft had crashed into a hill near Deal and it was later discovered that the pilot was Plt Off. Watling. He had joined 92 Squadron when it had been stationed at Pembrey and, as he was considered to be a good pilot, his loss was a severe blow to 92 Squadron.

On 11 February, the AOC of Fighter Command, Air Vice-Marshal Leigh Mallory, visited Manston and inspected the station. No. 92 Squadron flew a number of patrols, but there was no enemy action and the following day the unit carried out a sweep of the Channel, along with 74 Squadron from Biggin Hill and 66 Squadron from West Malling. That evening, the Blenheims of 59 Squadron carried out a successful raid, attacking objectives on the French coast and encountered no opposition.

No. 74 (Tiger) Squadron flew into Manston from Biggin Hill with their Spitfire IIA on 20 February, under the command of South African Squadron Leader Adolph 'Sailor' Malan. No. 92 Squadron moved out to Biggin Hill on the same day. Malan had joined the RAF in 1935 after a short spell in the Navy, hence his nickname 'Sailor'. He had taken over the Squadron on 8 August 1940 and was known for his discipline and no-nonsense approach. Under his command, the unit was one of the first to abandon the Vic formation, which had been much criticised, and adopted the 'finger-four' that was being used by the Luftwaffe. It consisted of two pairs of fighters flying together, with the second pair positioned above and behind the first couple to provide top cover.

It was a bright, clear day with visibility of 12 to 15 miles, when, at 12.45 p.m. on 22 February, approximately fifteen Bf 109s attacked the airfield, while another dozen or so fighters flew above them as top cover. The aircraft dropped their bombs from estimated heights of between 8,000–10,000 feet, but a number of them ventured below 5,000 feet and strafed various targets on the airfield. Fifteen bombs were dropped in less than five minutes, but just a single unlucky airman was killed while he was making his way to an air raid shelter. Eight bombs were dropped on the landing ground, but only one aircraft was damaged; on the domestic site, the decontamination centre and three barrack rooms were totally destroyed.

Just four days after arriving at Manston, 74 Squadron suffered its first losses, when Sergeants N. Morrison and Polish pilot Sergeant Rogowski failed to return from a convoy patrol. Rogowski had made a crash-landing near Eastbourne, flying Spitfire II P7559, and was only injured, but neither Morrison's aircraft, P7618, nor his body were ever found.

The German raids continued and, at 7.35 a.m. on 9 March, twelve-plus Bf 109s were observed circling the airfield and a number of them dropped their 50-kg bombs from an estimated height of between 5,000–8,000 feet. At least two other pilots were more adventurous and went down to around 2,000 feet to bomb and shoot up the airfield. A total of nineteen bombs were dropped, but fortunately the only damage was a large number of broken windows and one wounded airman.

A Dornier 215 also participated in this raid and its crew may have been observing or even co-ordinating the attacks by the fighters. The Dornier did not attempt to attack, but circled the airfield at approximately 10,000 feet and was engaged by the AA battery with 3-inch guns, but no results of it being hit or damaged were observed. There was another raid by a single enemy aircraft at 2.50 a.m. on the 14th, when two 250-kg bombs were dropped on the airfield; however, there was no damage.

The command of 74 Squadron changed on 10 March, when Squadron Leader John Colin Mungo Park, DFC, took over from Squadron Leader Malan, who was promoted to Wing Leader. Park was a lot easier to get on with than Malan and was occasionally known to turn a blind eye to certain things. Having been made a Flight Commander in September 1940, Park had been promoted to acting squadron leader in November and was awarded the DFC.

Wing Commander Graham Ashley Leonard Manton was replaced as the Commanding Officer of RAF Manston in March, having held the post since September 1940. Manton, the former Commanding Officer of 56 Squadron, was appointed as Wing Leader at Northolt. Manton's replacement was Wing Commander R. L. Bennet, who had spent some time at Acklington as the Senior Administrative Officer. He has been described as an extremely ambitious officer with his aim being to achieve 'air rank' as soon as possible. Soon after his arrival, he had also read the Riot Act to the pilots of 92 Squadron, which did nothing for him in the popularity stakes either.

April was an eventful time, although it began quiet enough—on the first day of the month, there was no flying by the resident units. There was no flying by resident units the following day either, but, during the evening, two Spitfires IIAs from 91 Squadron (91 Squadron had recently moved to Hawkinge from Kenley) landed at Manston after carrying out an uneventful Jim Crow sortie (a typical Jim Crow sortie was a patrol of the English Channel to carry out general reconnaissance for Fighter Command). Both Spitfires flew back to Hawkinge early the following morning.

The RAF, and Fighter Command in particular, was now on the offensive and Manston was to play a significant role in the new types of sorties that were meant to do to the Germans what the Luftwaffe had been doing to the RAF for many months. Sorties known as Rhubarbs were low-level offensive patrols carried out by two or more aircraft that attacked targets of opportunity, such as trains, barges, or road convoys. Roadstead sorties were attacks on the enemy's coastal shipping; Rodeos were offensive fighter sweeps, and the first one had taken place on 9 January. Circus sorties were bombing raids by light bombers such as Blenheims and Hampdens that were protected by a large number of fighters. Intruder sorties were normally carried out by a single aircraft and were meant to penetrate enemy defences at night and bomb or strafe enemy airfields.

At 4.30 p.m. on 6 April, two Spitfires from 74 Squadron took off from Manston on a Rhubarb sortie in the area of St Omer; Plt Off. Bob Spurdle (754228), who came from New Zealand, attacked and damaged a Bf 110 after it had attacked and damaged his aircraft. Flt Cdr Flt Lt Tony Bartley was the pilot of the other Spitfire and he claimed to have attacked and downed a Bf 109 that had made a belly-landing on one of enemy airfields they had flown over. Spurdle returned to Manston and landed safely at 5.15 p.m., but, when he climbed out of the aircraft, he realised that he had been very, very lucky—in his flying jacket he found the fragmented remains of a cannon shell that had failed to explode. Flt Lt Bartley's Spitfire had also been damaged in the action and he had made a belly-landing at Hawkinge at 5.40 p.m.

The following night, 23 Squadron, which had a detachment at Manston, sent a Blenheim to Vitry-en-Artois on a night Intruder sortie flown by Plt Off. Robinson. The Blenheim took off from Manston at 10.20 p.m. Robinson arrived over his objective at 11.03 p.m. and the airfield was clearly visible in the moonlight, but there was no sign of any aircraft on the ground. Robinson circled the airfield for over twenty minutes, but he was not challenged by either the ground defences or enemy aircraft and at 11.29 p.m. he dropped a stick of forty 40-lb bombs, which were seen to explode on the runway. From Vitry-en-Artois, he proceeded to the airfield at Lille-Sud, where all the lights on the airfield were suddenly extinguished; he safely returned to Manston and landed at 12.30 a.m.

There was no operational flying during the day of the 9th, but the Blenheims of 23 Squadron were active again that night and Fg Off. Davies and his crew were the first to take-off at 11.35 p.m. to attack the airfield at Merville, where they dropped eight 40-lb bombs. They then proceeded to Lille, but after observing no activity they returned to Merville, where they saw a fire burning on the eastern boundary. Plt Off. Love and his crew were airborne at 2.25 a.m.; they flew over Merville, which appeared inactive, and then they flew to Béthune, where they reported seeing a long straight line of lights interspersed with arrows and crossbars. This was claimed to be the first sighting of the Luftwaffe's 'Visual Lorenz' lighting system that had become the standard lighting pattern on German airfields.

The next day, there was a tragic accident at Manston that could so easily have been avoided; it involved a 74 Squadron pilot, twenty-year-old Plt Off. Peter Chesters. Chesters, the son of William and Kate from Thorpe Bay, Essex, had joined the Royal Air Force Volunteer Reserve in 1939 and had trained to fly at 5 SEFTS (Service Elementary Flying Training School). before moving on to 7 Operational Training Unit at Hawarden, Chester. On the completion of his training, he requested that he be posted to 74 Squadron and he was subsequently posted to that unit on 29 September 1939.

At 4.45 p.m. on 10 April, six Spitfires from 74 Squadron took off from Manston to patrol Folkestone at 25,000 feet, and within a short while they encountered a number of Bf 109s. The Bf 109s, from JG51, were dispersed and one of them was shot down by Plt Off. Chesters, who then returned to Manston and attempted to carry out a 'victory roll'. Unfortunately, he was either too low or he mistimed his actions, failed to recover, crashed on the barracks square, and was killed instantly.

Plt Off. Chesters was a well-respected and experienced pilot, who had flown his first operational sortie on 2 October 1940 and claimed his first enemy aircraft on the 27th of that month. On that occasion, Chesters ended up in another kind of fight after he had followed an enemy aircraft down and landed on the same airfield to help the German pilot, *Feldwebel* Lothar Schievehofe, from the cockpit. However, the German refused Chesters' help and is said to have spat into his face causing a fight to break out that was only stopped when a policeman and ARP warden arrived on the scene. The policeman took back the Iron Cross that Chesters had torn from around Schievhofe's neck as a souvenir and he had to be content with the Bf 109's first aid kit.

On 27 November 1940, Chesters had been flying Spitfire, P7306, on patrol with 74 Squadron at 25,000 feet above Chatham, when he had been shot down by a Bf 109 of JG51 and forced to bale out over Blackett's Marsh, Essex, landing in an area known as Conyer's Creek. The weight of his flying gear pushed him into the mud and he was unable to free himself. Fortunately, a man who worked at a local brickworks had witnessed the whole incident and rushed to help Chesters. He had also been wounded in the leg and spent several weeks in hospital; he did not returned to flying until 17 January 1941.

There was also an incident on the 10 April involving a Blenheim, V5596, of 82 Squadron that had been attached to Manston from its base at Bodney in Norfolk. It had taken off from Manston at 5.35 p.m. to bomb and strafe objectives at Borkum, where the crew claimed to have caused considerable damage shooting up troops and attacking an E-boat. During the return flight, the aircraft was fired at by a flak battery at Ostend and as a result the pilot became disorientated, lost, and confused by repeated sightings of the coast. Subsequent events could be described as farcical.

With his aircraft damaged and short of fuel, the pilot was forced to put the Blenheim down on a sandbank, but unfortunately he and his crew were under the impression that they had landed on the wrong side of the English Channel. As per orders, the crew quickly organised the destruction of all the classified documents and equipment. They were in the process of launching the dingy to escape across the Channel when the Margate lifeboat arrived on the scene. One can only imagine how they felt when they realised where they were and that they were just a short distance from the safety of the airfield at Manston.

On 11 April, four Defiants were diverted to Manston from Biggin Hill and Gravesend, but two of them overshot the runway and were damaged. The other two left during the early hours of the following morning. By now, the Defiants were operating in the night-fighting role as an interim measure until the Bristol Beaufigter became fully operational.

No. 74 Squadron flew out to Gravesend on 1 May and was replaced by the Hurricane IIs of 601 (City of London) Squadron that flew in from Northolt. No. 601 Squadron was commanded by Sqn Ldr E. J. Gracie DFC, who had recently been posted in from 23 Squadron. The Squadron's first patrol was at 3.30 p.m. the same day, when one of its Hurricanes patrolled the Thames Estuary, landing at 4.45 p.m. with nothing to report.

2

The Flying Dutchmen

A highly unusual event occurred on the morning of 6 May, when, at 7.45 a.m., a Fokker T.VIII floatplane was seen approaching Viking Bay near Broadstairs. It was no surprise when anti-aircraft guns opened fire due to its evidently German markings. The Fokker made a bumpy landing in shallow water near the beach where the engines were turned off, but the pilot started the engines again and taxied another 300 yards across the bay and moored by the pier. Soldiers from the Home Guard cautiously approached the aircraft with their rifles at the ready.

The Fokker's four-man crew appeared waving white handkerchiefs and what everyone thought was a German flag, but Richard Skeels, a sailor attached to HMS *Fervent*, a land base in Ramsgate, was among those who spotted something strange—the airmen were wearing civilian clothing and seemed to have broad grins on their faces.

As the four airmen raised their hands in the air and shouted 'don't shoot! We are Dutch!', Skeels and others on the scene soon realised that the airmen were not Germans and began to treat them more cordially. They were taken to the local police station where they were given a hearty breakfast before being collected by a couple of RAF officers and driven to RAF Manston. During the debrief, they began to tell their story, which was one of the great escapades of the Second World War; they had stolen the Fokker seaplane from under the noses of the Germans and escaped from occupied Holland.

The plan to steal an aircraft and escape to England had been mulling for a while and originally it had involved a plot to steal a Ju 52 from Schipol Airport with the help of a KLM pilot. That idea had to be abandoned, but when Lt Govert Steen, a demobilised Dutch fighter pilot, managed to get work at the Fokker factory, a new one was hatched. He became aware that the Germans had been testing a floatplane, a Fokker T.VIII, which was sitting on the water in Minerva Harbour, Amsterdam. The aircraft was a twin-engined torpedo bomber that had been designed in 1938. Being a pilot and having knowledge of the type of aircraft made Steen a vital part of any escape plan.

One of the other men on the aircraft was Evert Willem Boomsma, who was being trained as an aircraft mechanic when the Germans had invaded Holland in May 1940. His father was involved with the Dutch Resistance and not long before he escaped to England, he had found out that his father had been arrested and that the Germans were looking for him as well. The fear of being betrayed played on his mind and with the Gestapo closing in on him he had nothing to lose.

When Boomsma and Steen heard that two other men who they knew, Vos and Leegstra, had escaped from Holland in a stolen Fokker G.I on 5 May, they were aware that they had to act immediately, before the Fokker T.VIII was flown out and the Germans tightened their security. There were a number of things to sort out before any escape plan could be put into action, such as finding an inflatable dinghy or small boat to get out the floatplane. They also needed the right weather conditions, with a westerly wind to take-off in the harbour.

The other two men aboard the Fokker were Dutch Army Lieutenant Beelaerts Van Blokland and technician Wiebert Lindeman. It was decided that, so as not to arouse too much suspicion, the four men should go to the harbour in pairs. Steen and Blokland caught the first tram and Lindeman and Boomsma got on the next one. They had found an inflatable dinghy and had broken it down into four pieces, with each man carrying one. Boomsma was also carrying a Dutch flag, which he rightly assumed would come in handy when they landed.

After a scare when the four men had bumped into a cyclist who they thought might give them away, at 2 a.m. they returned to the bridge that they had chosen as their meeting point and began to blow up the dinghy with a pair of bellows. They lay in the grass for most of the night, waiting for first light when they would be able see enough to manoeuvre the aircraft. Eventually the dingy was lowered into the water and with the help of a single paddle, the four men cautiously began to move towards the Fokker. When they arrived at the aircraft, which carried the code letters 'KD+CQ', they removed the engine covers, but were frustrated waiting for dawn to break.

Wiebert Lindeman, who was to manage the fuel system and flaps, began to familiarise himself with the engine controls. Beelaerts was positioned in the bomb aimer's position and Boomsma chose to sit in the back of the aircraft. Meanwhile, Steen sat in the pilot's seat, trying to work out the dials and instruments. With Lindman, Steen was soon able to give the others some very good news because he had found that the aircraft's fuel tanks were full to the brink.

At 7 a.m., when Steen was about to start the engines, an Arado 196 floatplane flew over the top of the Fokker and he was momentarily startled; he quickly fired the port engine, which burst into life immediately, and he throttled it back while Boomsma untied the strap from the float to release the aircraft from its anchor, at which point Steen started the starboard engine. As the starboard engine burst into life, the port engine suddenly died and the Fokker began to turn in circles on the water, with one float still tied to the anchor.

To the four Dutchmen, it seemed ages before the port engine started again, but it was probably no more than a minute or so. Boomsma released the other rope on the float and then climbed back in, as Steen began to taxi the aircraft out of the harbour. With Lindeman controlling the throttles, they slowly moved on to the Noordzee Canal that connects Amsterdam with the North Sea. They taxied right past a German guard post, but nobody took any notice, suggesting that they thought it was perfectly normal for the aircraft to be about to take-off.

As they gathered speed across the water, the Fokker bounced along, but refused to get airborne, while ahead of them was the Hem Railway bridge that presented the final obstacle in their path. It was a close-run thing, but the aircraft just about cleared the bridge, with Boomsa later claiming that they probably left a coat of paint on it.

With Steen still getting used to the Fokker's controls, he flew the aircraft in the direction of Zandvoort, where they would begin to cross the North Sea towards England. It was at that point that Steen realised that the aircraft did not have a compass and all he could do was to point the aircraft in what he thought was the right direction and hope for the best.

To begin with, Steen managed to climb up into the cloud so as to hide the aircraft from any unwanted attention, but the cloud got thinner the further out to sea that they flew and then disappeared completely. To compensate for that, Steen decided to fly as low as possible, just above the waves and so low that it made the others nervous. After about one hour and fifteen minutes, they spotted what they were sure was the English coast, but they did not have any idea where they were. It was at that point that Steen turned the aircraft into wind and flew parallel with coast, displaying the German marking that drew the attention of the defences.

After being interrogated at Manston, the four Dutchmen were taken back to Broadstairs Police Station, but later that day they were driven up to London and questioned again at the Dutch Patriotic School, where a surprise awaited them. As they walked in to the room, they met Vos and Leegstra, the two men who had escaped from Holland in the Fokker G.1 the day before. There was a great deal to catch up on and jokes abounded about the Germans being so careless as to lose two aircraft in two days. However, the British authorities were still somewhat suspicious about recent events. Although some aircraft had been evacuated from Holland and flown to England in the days immediately after the German invasion, since that time not a single similar escape attempt had been made, but now there had been two on consecutive days.

On Saturday 10 May, exactly one year after the Germans had invaded Holland, it had been arranged that the four escapees meet Prince Bernhardt of the Netherlands. Unfortunately, the Prince was attending a Service of Remembrance and was unable to meet them, although he did arrange it for the following day and they had an informal meeting. On the 14th, they met Queen Wilhelmina and discovered that the story of their escape had been broadcast on Radio Oranje in a coded message and they were happy for everyone in Holland to know what they had done.

The aircraft that Steen and his crew had flown over from Holland was flown to be stored at Felixstowe by Fg Off. Schaper, who had flown another Fokker T.VIII to England on 14 May 1940. That aircraft had been one of eight of the type that had been flown to England on the day when the remnants of the Dutch Air Force was evacuated.

The Fokkers were allocated to 320 (Dutch) Squadron, which had been formed at Pembroke Docks on 1 June 1940, and remained in service with the RAF for a year and half, mainly flying stealth operations into Holland. Many of them were broken up for spare parts when they were replaced by the Avro Anson.

Govert Steen joined the RAF and was later posted to 129 Squadron, which was equipped with Spitfires, but he was killed on 5 June 1942, flying BM639 and escorting Bostons during a raid over France. Van Blokland and Lindeman joined the Princess Irene Brigade, made up of ex-Dutch nationals. Blokland went on to command the unit, which later became part of the 21st Army Group. Boomsma, who had escaped to England to avoid capture by the Gestapo, also served with the RAF in 320 Squadron, but returned to Holland in 1951. His father, Reinder, who had been arrested a short while before he flew to England, died in Neuengamme Concentration Camp on 26 May 1943.

3

Channel Stop

During its time at Manston, 601 Squadron flew numerous sorties against shipping as part of the 'Channel Stop' operation, but the comment 'nothing to report' was a regular phrase that appeared alongside its sorties in the Manston ORB. On 6 May, two of 601 Squadron's Hurricanes took off on a training patrol, but were attacked by a Spitfire, the pilot of which must have identified them at the last moment and pulled away without firing. Then the same two Hurricanes were attacked by two Bf 109s, but a flight of Spitfires, including the one that had just threatened to shoot them down, appeared on the scene. One of the enemy aircraft quickly disappeared into cloud, while the other one was hit and damaged, but managed to pull away from the fight.

The Boulton Paul Defiant did not have very good reputation, even after it had been relegated to night-fighting duties, but it did fill a gap until the Beaufighter entered service. During the night of 8 May, a Defiant of 264 Squadron was positioned at Manston from West Malling and it flew an offensive patrol over Merville Aerodrome in France, where Fg Off. Young and his gunner, Sgt Russell, claimed to have destroyed a Bf 110 on the outskirts of Caen.

Merville was visited again on the 11th by Sqn Ldr Gracie, the CO of 601 Squadron, who, having served on 23 Squadron, was said to be an experienced pilot when it came to Intruder sorties. When he arrived over Merville, the aerodrome lights were on, then suddenly two white flares were fired, but they were quickly extinguished. Gracie then flew to Lille-Sud aerodrome, where he encountered an enemy aircraft in the circuit with its lights on and about to land. The aircraft doused its lights, flew across the middle of the airfield, and then put its lights back on again with Gracie giving chase, but then, fearing a trap, he called the operation off and returned to Manston at 3 a.m.

No. 601 Squadron had some success on 16 May when two Hurricanes took-off at 2.25 p.m. to patrol between North Foreland and Dungeness and Sergeant McCann spotted a Bf 109 heading north. MacCann managed to get in a short burst while flying head on towards the enemy aircraft, but then it climbed to pass over his Hurricane and headed back to France. Czech pilot Sgt Mares then got in a three-second burst before McCann swung around and got in another three-second burst, which caused the BF 109 to crash into the sea 10 miles south of Dover.

Being on the coast, both Manston and Hawkinge were popular airfields that pilots would make for when they were in trouble with a damaged aircraft. Sgt Chestnut of 609 Squadron, based at Biggin Hill, was trying to land at Manston on 11 June after his Spitfire, P8654, had been badly shot up in combat with a Bf 109, flown by *Oberleutnant* Ebersberger from JG26. At 5.05 p.m., the Spitfire crashed into the cliffs near Ramsgate, and Chestnut had no chance at all of surviving the impact. Sgt Guy Alexander Chestnut (R61465) was a Canadian from Saskatchewan and he was buried with full military honours in Margate St John's Cemetery.

The Goodwin Sands was also a popular spot, where pilots could make a forced-landing with a reasonable chance of surviving and being picked up. Fg Off. D. Stewart-Clarke from 603 Squadron crash-landed Spitfire R7345 on the Goodwin Sands on 21 June after an engagement with a Bf 109 of JG26, in which he was wounded. It is not an exaggeration to say that the Goodwin Sands are littered with aircraft wrecks, one of the most well-known being Avro Lancaster, JB278, formerly of 103 Squadron.

The advantage of pilots trying to reach the safety of land was that sometimes the aircraft could be repaired and put back into service, although for a pilot that was wounded this was not always an option. On 23 June, Plt Off. B. M. Gladych of 303 Squadron made a crash-landing near Ramsgate in Spitfire, P8338, after colliding with a Bf 109. The aircraft was repaired and served on until it was struck-off charge in March 1945. Sergeant Lamb of 603 Squadron made a forced-landing near Walmer in Spitfire W3364 on 24 June. This aircraft was also repaired and put back into service.

The Hurricanes of 601 Squadron were replaced by the Spitfires of 222 Squadron at the beginning of July, and it was the usual straight exchange with the Hurricanes replacing the Spitfires at Matlask in Norfolk. On 19 July, 242 Squadron flew in from North Weald to replace 222 Squadron, which was posted out to Southend. No. 242 Squadron was equipped with the Hurricane IIB, the armament of which was a powerful mix of twelve machine guns, rockets, and bombs. Their role was to silence the flak thrown at them and allow other aircraft such as Blenheims to attack German shipping.

No. 242 Squadron was kept busy during August, flying on Roadstead and other offensive patrols, but they were to prove costly in terms both airmen's lives and aircraft. On the 17th, 242 Squadron flew a Roadstead between 6.55 p.m. and 8.05 p.m., claiming a destroyed flak-ship and a single BF 109. It also lost two Hurricanes, Z3845 flown by Plt Off. K. M. Hicks and Z3454 flown by Plt Off. E. A. Redfern were both shot down by a Bf 109.

Sergeant Guy Alexander Chestnut of 609 Squadron, whose Spitfire crashed into the cliffs at Ramsgate on 11 June 1941. He is buried in Margate's St John's cemetery. (*John Williams*)

The headstone of Sergeant Chestnut in Margate Cemetery. (*John Williams*)

No. 615 Squadron, under the command of Sqn Ldr Gillam, relieved 242 Squadron on 10 September, when that unit moved north to Valley on the Isle of Anglesey. No. 615 Squadron flew its first sortie from Manston on the 14th, when one of its Hurricanes went on a shipping reconnaissance operation and claimed to have destroyed an armed barge. During a sortie on 26 September, a 615 Squadron Hurricane destroyed two enemy mine sweepers, a flak-ship, and another vessel described as a small destroyer. The following table is taken from the ORB, September 1941:

Summary of Shipping Operations during First Nine Months of 1941

Month	Ships Over 1,000 Tons	Others
January–June	31	Many
July	18	Many
August	7	Many (Mine Sweepers)
September	8	

All Sunk or Severely Damaged

Enemy Losses	Destroyed		Probably Destroyed	
	Ships	Aircraft	Ships	Aircraft
By RAF Bombers	42,500 tons	0	26,000 tons	0
By RAF Fighters	2,100 tons	23	1,500 tons	11

RAF Losses
Bomber Crews: 13
Fighter Pilots: 20

The Importance of Manston up to the end of September can be gauged from the figures given above, since most of the attacks concerned were by Manston based squadrons. It is impossible to over-estimate the significance of the work done by 242 and 615 Squadrons. They virtually closed the Channel to enemy traffic by daylight.

Within a matter of months that last sentence might have come back to haunt the officer who wrote this report.

Wg Cdr Thomas Percy Gleave arrived at Manston in October and took over as the Commanding Officer on the 3rd, from Wing Commander Bennett. Having been born and educated in Liverpool, Gleave had become interested in aviation at an early age and he had gained his private pilot's license in 1928. He then went to Canada, where he had worked in a tannery, but in 1930 he returned home and was commissioned into the RAF. By 1933, he had become a member of the RAF's aerobatic team and had trained as a flying instructor before joining Bomber Command in 1939.

Gleave's time in Bomber Command was short and by 1940 he was back in Fighter Command and had become the Commanding Officer of 253 Squadron, which had been one of the first fighter units to arrive at Manston after the outbreak of war. Sqn Ldr Gleave had flown in the Battle of Britain and had been credited with five enemy aircraft before he himself had been shot down and badly burned on 31 August 1940. With his clothes on fire and the skin on his hands beginning to blister, Gleave had to bale out, but his oxygen tube was jammed. It was only when the aircraft exploded that he was ejected from the aircraft.

His parachute opened and he came down in the vicinity of the airfield at Biggin Hill, where he was taken to Orpington Hospital and immediately operated on. In the Royal Victoria Hospital, he became a patient of Archie McIndoe, who gave him a new nose; while he was still in hospital, he was promoted to wing commander. It was not until October 1940 that Gleave had been declared fit enough to fly on operations again and, after a brief spell serving as the Commanding Officer of Northolt, he was posted to Manston. Probably because of his own experiences, Wg Cdr Gleave became one of the station's most popular and renowned Commanding Officers ever to serve there.

The second detachment of 607 Squadron arrived at Manston on 13 October and, although there is no mention of the first detachment arriving in the ORB, it is generally accepted that it had landed on the 10th. The unit would work closely with 615 Squadron on anti-shipping operations.

The Secretary of State for Air, Sir Archibald Sinclair, visited Manston on 18 October. Sinclair, who had been a major in the Guards Machine Gun Regiment during the First World War, had been appointed to the office on 11 May 1940. Although he was to keep the position until the end of the war, he did not always agree with Prime Minister Winston Churchill and he had his own ideas on how things should be done.

Manston had another distinguished visitor on 29 October, when General De Gaulle arrived to meet and talk to French aircrew, particularly those pilots serving with 615 Squadron. Photographs taken at the time show him out on the airfield talking to a number of Spitfire pilots, also meeting with Air Marshal Leigh Mallory, the AOC of Fighter Command, and Air Marshal Hugh Dowding. De Gaulle, who had been condemned to death by the Vichy government for treason in August 1940, had recently formed the Free French National Council in London.

It was a bad day for Manston on 2 November, when three pilots along with three Hurricanes were lost in action. At 5.10 p.m., two Hurribombers of 607 Squadron had taken off with two Hurricanes from 615 Squadron on a

Future French President, Brigadier General Charles André Joseph Marie de Gaulle visiting RAF Manston on 29 October 1941, meeting French pilots of 615 Squadron.

A more informal pose of de Gaulle outside what appears to be the old officers' mess in Pouce's Farm building with the Commander in Chief of Fighter Command, Air Marshal Sholto Douglas.

de Gaulle taking the salute and shaking hands with an unidentified air marshal.

shipping reconnaissance. A couple of minutes later, a further two Hurricanes from 607 and 615 Squadrons took off on a shipping reconnaissance to Ostend. It was noted that a single Hurricane had landed back at Manston at 6.20 p.m., but that three pilots and aircraft were missing.

The Manston ORB does not state which aircraft or airmen were lost, however, we now know that they were the two Hurricanes from 615 Squadron and a single Hurricane from 607 Squadron. The 615 Squadron pilots were FS J. E. Slade in BE144 and FS A. E. Gooderham in Z3841, who were shot down by flak off Ostend. Gooderham was an experienced pilot and had been shot down before on 15 October 1940 while serving with 46 Squadron, when he had baled out over the Thames after a combat with a Bf 109.

The Hurricane of 607 Squadron was flown by Sgt W. C. Lees and was also shot down by flak. No. 607 Squadron's attrition rate was quite high and on the 4th it lost another Hurricane, which was shot down while attacking objectives at Le Touquet, although its pilot, Sqn Ldr G. D. Craig, survived to become a POW.

No. 217 Squadron, equipped with Bristol Beauforts and based at Thorney Island, had regular detachments at Manston. On 10 November, one of its aircraft crashed half-a-mile east of the airfield while on a training flight. Beaufort, serial number L9971, was being flown by Sgt P. L. Lankin, and both he and the other three members of the aircraft's crew were killed instantly.

A report written by the CO of 615 Squadron, Sqn Ldr Denys Gillam, dated 11 December 1941, outlined the experience gained by 615 Squadron during its operational tour at Manston. The unit had flown its final patrol

A Bristol Beaufort of 217 Squadron. Developed from the Blenheim, it was the standard torpedo bomber used by Coastal Command. Several of this type of aircraft took part in the Channel Dash on 12 February 1942.

Flight Lieutenant Finch of 217 Squadron, who took part in the attacks on the SS *Madrid*, was killed during the Channel Dash operation.

at Manston on 30 November, when it had been replaced by 32 Squadron. It was entitled *Tactical Memorandum No. 11*, and it stated that the operational sorties carried out were divided into two classes:

1. Silencing flak during shipping attack.
2. Silencing flak during mass low level attacks, i.e. Ramrod Operations.

On arrival at Manston, it was found that the enemy were running convoys through the Straits in daylight, approximately one every three days. The convoys consisted of one or two ships of 2,000 to 6,000 tons, and nine to fourteen flak ships of varying size and at least one destroyer. To get over the withering flak, it was found necessary to carry out synchronised attacks with as many aircraft as possible, in order to confuse the aim of the gunners and prevent them from singling out any one aircraft.

In order to give a complete picture, it is necessary to describe one particular operation in detail.

Assuming that a convoy has been located and that its size is fairly big, the following forces would carry out an attack:

8 or 12 (615 Sqn) Anti-flak Hurricanes.
8 Hurricane bombers or 3 Blenheims.
1 to 3 escort Spitfires

The bomber and anti-flak squadrons would form up together on Manston Aerodrome and await the arrival of the Spitfire escort. The Spitfire escort would arrive under 500 feet and do one circuit of the aerodrome. Meanwhile, the Hurricanes on the ground would take off and set course immediately at sea level. The following formation would then be adopted:

615 Squadron leading in pairs in line abreast with a gap of approximately 200 yards between the Flights.
The bomber Hurricanes behind the line abreast also, and with the Spitfire squadrons or Flights on either flank.

The whole formation would cover a front of up to a mile with a depth of only 200 yards and would cross the enemy coast at sea level and then turn up or down the coast just out of range of the shore guns.

On sighting the coast, the Spitfire escort would pull up to approximately 2,000 feet above the main attacking formation.

On sighting the enemy convoy, an order on the R/T of 'Buster' would be given by the leader. On receipt of this order, the attacking force would open up the throttles and climb steeply—the outside sections going ahead up to 1,500 feet; the inner sections up to 1,000 feet; the bombers up to 800 feet. Then reaching a position where the whole formation is embracing the convoy, an order to attack would be given. On this order, all the attacking aircraft would dive immediately to attack; the position before the attack should ensure that the attack is delivered from all direction.

While most pilots were happy to carry out such detailed instructions, there were a small number of others who, regardless of the purpose of the sortie, preferred to operate alone.

4

The Lone Wolf

On 24 November 1941, three Hurricanes arrived at Manston to form a night-fighter intruder flight. The pilots were Flt Lt Stevens, who was attached from 253 Squadron, with Flt Lt Crabb and Sgt Scott of 3 Squadron. Soon afterwards, the flight was reinforced with the arrival of Flt Lt Shaw and Sgt Gilbert.

Thirty-two-year-old Flt Lt Richard Playne Stevens was something of a legend among night-fighter pilots because he was not only one of the oldest, but one of the most experienced, having flown as a commercial pilot with Imperial Airways. Born in Tunbridge Wells in September 1909 and educated at Hurstpierpoint College, Sussex, Stevens was one of six children, having four brothers and a sister. In his youth, Stevens had travelled out to Australia, where he worked on a cattle ranch, but he later returned to Britain and in 1932 he was married to Olive Mabel. He had then travelled overseas again to the Middle East and became a member of the Palestinian Police Force before returning to Britain again in 1936. It was only then that Stevens began his flying career, taking flying lessons at Shoreham Flying School.

His first job in aviation was with Imperial Airways, flying as a co-pilot on the Croydon to Paris route, often flying at night and gaining a relatively large amount of experience—approximately 400 hours. It is thought that Stevens joined the Royal Auxiliary Air Force in July 1937 and was formally accepted in April 1940. By July that year, Stevens was holding the rank of sergeant and was stationed at Number 6 Anti-Aircraft Unit, Ringway Airport, Manchester, and his name appears in the logbook of observer/air gunner Sgt G. E. Wainright (651749). The first entry was on 12 July, when Sgt Stevens flew a Percival Q6 (G-ADDX) to Oswestry for practise dive bombing on a gun battery. Wainright made two further flights with Stevens that are mentioned in his book, the last one being on the 20th, when they landed at Cardiff with radio problems.

As to when Sgt Stevens was posted out of 6 AACU at Ringway is not known, but by October 1940 he was serving on 151 Squadron, which was based at

Digby in Lincolnshire, although it was later moved to Bramcote and then Wittering. The unit was a night-fighter squadron that was initially equipped with the Hurricane, but also had a number of Boulton Paul Defiants. No. 151 Squadron was then commanded by Sqn Ldr West, who had taken over from Sqn Ldr King after he had been killed. Around about the time that Stevens joined 151 Squadron, he suffered a personal tragedy that may have affected him in all aspects of his life.

There were rumours that Flt Lt Stevens had been badly affected psychologically after his wife had been killed during the Blitz in Manchester. This incident was said to have brought on a complete hatred of Germans and caused him to act rashly in his quest to destroy as many aircraft and kill as many pilots as he could. However, recent information suggests that it was not his wife that had been killed but his daughter, and she died as a result of a fire caused by a German air raid during the Blitz.

Whatever happened in the air, Stevens was not by nature a violent or aggressive person, but quiet and contemplative. In the crew room, he was said to have been a thoughtful man and was often to be found slumped in an armchair reading the works of T. E. Lawrence.

The first enemy aircraft that Stevens destroyed was on the night of 15–16 January 1941, when, after taking off from Wittering in Hurricane V6934 and after flying through the London barrage, recently commissioned Plt Off. Stevens gave chase to a Do 215. Although the pilot dived to escape and the gunners returned fire, Stevens doggedly pursued the Dornier and shot it down. Within a short while, he found another enemy aircraft, a Heinkel, over west London, which he also chased and shot down into the sea off Canvey Island. These were the first two enemy aircraft destroyed by 151 Squadron and one of only a very few occasions when a night-fighter had destroyed more than a single enemy aircraft in a night.

On 4 February 1941, Stevens' actions were recognised when he was awarded the DFC, and three months later came a Bar to the DFC. The reason why Stevens became a determined killer are not known, but he hated the Germans and adopted a ruthless approach towards his victims. On one occasion, it is claimed that he returned from a sortie with blood and the human remains of an enemy airman on the wing of his Hurricane, after the He 111 that he had been stalking exploded. The ground crew offered to clean it up, but Stevens told them to leave it alone.

The same afternoon that Stevens arrived at Manston, he was involved in the search for a pilot from 26 Squadron, whose Curtiss Tomahawk had been shot down 15 miles west of Ambleteuse. The unit, whose main role was Army co-operation, had been attached to Manston two days before on the 22nd. Flt Lt Stevens spotted the pilot after he had discharged a smoke float and guided a section of 615 Squadron Hurricanes to the scene. Green Section of

615 Squadron then provided cover for the air-sea rescue launch and all the aircraft involved had landed safely by 2.45 p.m.

It was also noted in the Manston ORB for 30 November that Flt Lts Stevens and Crabb and Sgt Scott had flown a number of uneventful night patrols. Stevens, who had only recently been posted to 253 Squadron, realised that to catch the enemy he would have to travel further afield, which is what he began to do.

On 12 December 1941, Stevens was informed that he had been awarded the DSO, but that did not keep him on the ground. At 7.33 p.m. that evening, Flt Lt Stevens took off and formated on a Havoc of 23 Squadron—quite a feat without the use of lights in blackout conditions. It seems quite likely that the Havoc was one that had taken off from Manston after Stevens' Hurricane.

Why Stevens did this is not known, but it may have been to make the crew of the Havoc more aware of the dangers of not keeping a sharp lookout. It is quite possible that Stevens followed and stalked the aircraft all the way as it flew towards its objective at Gilze-Rijen (Hulten) airfield. It was one of

Flight Lieutenant Richard Playne Stevens, the 'Lone Wolf'. A renowned night-fighter pilot, he was killed on 15 December 1941.

the oldest Dutch airfields and one of the Luftwaffe's most important bomber and night-fighter airfields. Having failed to make contact with the enemy that night, he returned safely.

Three nights later, on 15–16 December, Flt Lt Stevens took off at 7.40 p.m. on another sortie to the Dutch coast and Gilze via Overflakke, but his distinctive black-painted Hurricane failed to return. What exactly happened to Stevens is not known, but it has been suggested that he was stalking an enemy aircraft when he became distracted and flew into the ground.

Just 600 metres from where the wreckage of Stevens' Hurricane was found on the German-held airfield at Gilze-Rijen was the wreckage of a Ju 88 that had crashed at approximately 9.38 p.m. That was almost certainly Stevens' last victim, his fifteenth enemy aircraft destroyed. His body was at first buried at Breda (Zuijen), but it was later moved to Bergen op Zoom Cemetery where he lies in Grave 23.B.4.

One of the remarkable things about Flt Lt Stevens is that, during all the sorties that he flew at night, he was not aided by radar or other navigational equipment. He was what was known as a 'cat's-eyes' pilot and found the enemy only by using the Mk 1 Eyeball. Without the use of a navigator or controller to guide him, flying to the Dutch coast at night was no mean feat. On top of that, he then had to search for and stalk any potential victim and then find his own way back across the North Sea. That kind of flying called for precision navigation and only an experienced and skilled airman could carry out such operations.

Although largely forgotten by most military aviation historians, Stevens did receive the respect of his fellow pilots, including Johnny Johnson who mentioned him in his book *Wing Leader*. Johnson said that Stevens could always be found in the sky where the flak was the heaviest, prowling and searching for the hated enemy. He went on to say that he thought that his end was inevitable, and that he had the fondest memories of him.

5

Battling Beauforts

On 9 December 1941, three Bristol Beauforts of 217 Squadron were detached to Manston from their base at Thorney Island to attack a German ship that was moored close to the Dutch coast near Van Heldar. The objective was the 8,000-ton former passenger and cargo ship SS *Madrid*.

The vessel had been built in 1922 by the Vulcan Stettin Company and had for many years plied the Hamburg to South America route. It had originally been called the *Sierra Nevada*, but its name had been changed in 1937. The reason that the ship had been targeted was that it was being used by the Kriegsmarine as a U-boat tender that serviced, fuelled, and supplied them while they were at sea.

The operation was led by South African Flt Lt Finch and the other two pilots, Plt Offs A. Aldridge and M. Lee. When the three crews reported to the operations room at Manston, they were quite amazed when they found out that there was to be no fighter cover or diversionary tactics, just a straight 'in and out'. The three Beauforts took off from Manston at 3.42 p.m.

The Beauforts had been armed with four 500-lb bombs instead of torpedoes, and when Flt Lt Finch made his attack below mast height, it appeared to have been be successful. However, there was only a few seconds delay set on the fuses of the bombs and when they exploded the blast sent his aircraft hurtling up into the air. Next to attack was Plt Off. Mark Lee, flying in Beaufort AW190 (MW-K), but by the time that he made his run the German defences were prepared and the air was full of flak as his Beaufort dived towards the vessel. Within a few seconds, the port engine of Lee's Beaufort was hit and the aircraft burst into flames before crashing into the sea, killing all those on board.

The three other members of Plt Off. Lee's crew were twenty-nine-year-old Canadian FS John Ansley Foster and twenty-two-year-old Sgt John Alfred Chadway. There was also a Sgt John Henry Carter in the crew, but little is known about him. Sgt 'Harry' Carter was a married man with a family.

Plt Off. Arthur Aldridge was the pilot of the final Beaufort to attack and, despite having just witnessed the death of his best friend, Plt Off. Lee, he knew he had no option but to carry on. He thought that his best option was to fly low—very low—and so low that the flak guns could not be depressed far enough to hit him. As he flew above the ship, he felt his aircraft judder and for a few seconds he was convinced that he had been hit. During a quick glance around, he discovered that the end section of the port wing was missing and suspected that it had probably been sliced off by some bracing wire on the ship.

It was late in the afternoon by the time that Aldridge set course for home, but the controls felt heavy and he was not sure what other damage had been done and whether the Beaufort might just fall apart in the air. For a while, he pondered about the loss of his friend, Plt Off. Lee, who had been an all-round sportsman, but especially good at cricket, rugby, and hockey. During his second year at Cambridge University in 1940 and before he had been called up, he had begun flying training. Mark Lee was a bright, intelligent young man and, like many others, he had given up his life in the service of his country. He was later to learn that Plt Off. Mark Lees' body had been washed ashore at Ameland, a small island off the Dutch coast, and had been buried in Ameland Cemetery on the 12 December.

Darkness had fallen by the time that Plt Off. Alridge's Beaufort approached the airfield at Manston and his crew were unusually quiet, wrapped up in their own thoughts. He had doubts as to whether the undercarriage and flaps would be working and he was surprised when the wheels were lowered and everything else appeared normal, except for the navigation light on the port wing. After the Beaufort landed at 6.02 p.m., Plt Off. Aldridge was reluctant to tell the ground crew about the damage to the port wing on 'their aircraft', but they shrugged it off and said they would soon get it fixed.

No. 217 Squadron had already lost a Beaufort at Manston, when L9971 had crashed close to the airfield in November 1941 with the loss of pilot Sgt P. L. Ankin and three other members of his crew. On 5 January 1942, it lost another Beaufort, when AW285 crashed on the Canterbury Road and two of the crew were badly burnt, but survived, while Sgt Voy died of his injuries the following day.

Just a short while later, it came close to losing another aircraft and crew and in a spectacular fashion. The incident began when a Beaufort, flown by Fg Off. McGregor, was flying over Dover and he collided with a barrage balloon that was being flown to protect a convoy sailing through the Channel. McGregor had to use every bit of emergency power just to stay in the air and he immediately jettisoned the four 500-lb bombs the Beaufort was carrying.

When the Flying Control Officer at Manston heard what had happened, he came up with a plan for McGregor to land off an approach to the airfield from the direction of Ramsgate. It is quite likely, however, that he was not aware that the aircraft was towing over 1,000 feet of cable behind it. As the Beaufort flew low over the town, dragging the cable behind it, McGregor struggled to keep it in the air with the cable smashing roof tiles and demolishing chimneys along the way. When the Beaufort eventually landed, albeit in one piece, crowds of airmen and officers appeared, including the Station Commander, Wg Cdr Gleave.

When Aldridge had shut down the engines, the armourers jumped on board and one of them noticed that the toggle to release the bombs had been pulled and so he opened the bomb bay doors. At that point, everyone in the crowd fell silent in a state of shock as the four 500-lb bombs tumbled to the ground. With everyone fearing that they would explode at any moment, both officers and airmen ran from the scene as fast as their legs could carry them. Within a minute or so, the danger appeared to have passed as the bombs failed to explode and calm returned to the airfield.

What had happened was that, although McGregor had pulled the toggle to release the bombs over the sea, he had forgotten to open the bomb doors first and so they had been hanging loose. Along with the length of cable being dragged behind it, the bombs had been badly affecting the aircraft's trim and some considered it a miracle that McGregor had landed the Beaufort in one piece.

McGregor and his crew, along with a lot of other airmen and officers, who had stood around close to the aircraft, went off to the various messes to get drunk. Meanwhile, the ground crews cut the cable up into small pieces and all those who had been involved got a small piece of it as a souvenir of what might have been.

On 23 January, 217 Squadron detachment returned to Thorney Island, but the unit would soon return to Manston and be involved in another major operation that would lose more aircraft and crews. That was all the more reason for its airmen to appreciate those lighter moments when they got the chance.

6

Operation Fuller

On 9 December 1941, Manston had a number of American visitors in the form of general representatives of the United States Army Air Corps and the US Attaché, this being just two days after the Japanese attack on Pearl Harbor and the Americans declaring war on Japan. A few days later, there was another visit by Major Sparks from the US Army Air Corps, along with Sqn Ldr Berry from 3 Squadron. The Americans may have been looking for airfields that its air force could use and Manston certainly fitted the requirements.

There were another couple of high-profile visitors in January 1942, including the AOC of 11 Group Fighter Command, Air Vice-Marshal Leigh Mallory, who inspected all living quarters, squadrons, and sections. On the same day, WAAF officer Squadron Officer Corby Hall arrived to make arrangements for the WAAF contingent to be accommodated in Westgate.

The month of February began with anti-shipping sorties off Cape Gris Nez by five Hurricanes of 607 Squadrons that took off from Manston singly between 4 a.m. and 5 a.m. Only FS Hill and FS Moonaks spotted anything—three enemy vessels of approximately 5,000 tons—and they claimed to have damaged two of them as Category 4. All five Hurricanes had landed safely by 5.30 a.m.

There were no further operations for several days because of continuing bad weather in the form of low cloud, sleet, and snow showers, and it was not until the 6th that another two Hurricanes of 607 Squadron took off on convoy duty. On arriving over the convoy, the two pilots saw that one of the escort ships was firing at a Ju 88 that was threatening to attack. A single Hurricane gave chase and later landed at Martlesham having failed to catch the enemy aircraft. The other Hurricane landed safely back at Manston at 10.30 a.m.

On that occasion, the Ju 88 got away, but on the 14th another one was not so lucky after it had been spotted by two pilots of 607 Squadron heading north. They caught up with the Ju 88 and, during their attack, Fg Off. James and Sgt Blyth observed strikes on the fuselage and port engine. The enemy aircraft descended from 6,000 feet to 500 feet and both Hurricane pilots engaged in synchronised beam attacks, but it disappeared in cloud 5 miles from the French coast near Calais. Hornchurch operations later confirmed that the aircraft had crashed into the sea and James and Blyth shared the credit for one enemy aircraft destroyed.

February was a very busy and decisive month for the RAF, the Royal Navy, and 825 Squadron of the Fleet Air Arm, which was equipped with the Fairey Swordfish. The unit had been formed in October 1934 and had initially been equipped with the Fairey Seal. It had later embarked on HMS *Eagle* for China before returning to serve in the Mediterranean and Malta. No. 825 Squadron had been involved with the evacuation from Dunkirk and had lost eight of its twelve aircraft in March 1941; it had also been one of those units that had successfully attacked and sunk the *Bismarck* on 25 April 1941.

The Commanding Officer of 825 Squadron was thirty-two-year-old Lt Cdr Eugene Esmonde from Thurgoland, Wortley, near Barnsley, Yorkshire. He was the son of Doctor John and Eily Esmonde and had been educated at Wimbledon College and Clongowes Wood College in Ireland. He was another officer who had worked for Imperial Airways and had flown Class 'G' flying boats from Hythe to Malaya and Hong Kong. Esmonde had previously been appointed as Commander of 754 Squadron Fleet Air Arm, but on the 31 May 1940 he had been posted to command 825 Squadron (HMS *Kestrel*), taking over from Lt Cdr J. W. Hale.

No. 825 Squadron arrived at Manston on 6 February 1942, although there is no mention of that in the ORB. Not all of the unit's aircraft were sent to Manston and a detachment was also sent to Machrihanish in Scotland. The first mention of 825 Squadron at Manston was on 10 February, when a single Swordfish was sent to patrol the English Channel and the crew returned at 8.30 a.m. with nothing to report.

The German operation to move the battleships *Scharnhorst* and *Gneisenau*, along with the heavy cruiser *Prinz Eugen*, began during the afternoon of the following day on 11 February. The two battle cruisers had been a thorn in the side of the Admiralty since 1940, and had been responsible for the sinking of the aircraft carrier HMS *Glorious* on 8 June that year. German fighter ace *Oberst* Adolf Galland, Commander of JG26, had been surprised when called to attend a conference, along with the commanders of Luftwaffe units JG1 and JG2. They were informed of a daring plan that involved the three capital ships sailing from Brest to Germany through British waters and the English Channel in daylight.

The three capital ships were not sailing alone and escorting them would be six destroyers and fourteen torpedo boats, along with twenty-six E-boats. If that was not a powerful enough task force, the convoy would be protected by 176 bombers and 253 fighters, with which the Luftwaffe was to form an 'umbrella'. The Germans named their operation Cerberus after a mythological three-headed dog that guarded the gates to Hades.

On the 11 February, at 9.55 a.m., two Hurricanes of 607 Squadron had carried out a shipping reconnaissance patrol, but on landing the pilots had nothing to report. Between 6.15 p.m. and 10.05 p.m. that night, a Swordfish of 825 Squadron carried out two separate patrols in the Dunkirk–Ostend area, but also found nothing unusual to report. By the time that the Swordfish landed, the warships of the Kriegsmarine had already been at sea for nearly an hour, having left Brest at 9.15 a.m. It would be over twelve hours before the German fleet was detected and in that time the Kriegsmarine gained a great advantage over the British Navy and RAF.

At 10.20 a.m., two Spitfires from 91 Squadron took off from Hawkinge, flown by Sqn Ldr Oxspring and Sgt Beaumont. They were taking part in a Jim Crow patrol to see what was going on over the other side of the Channel. Some 15 miles west of Le Touquet, the two pilots found themselves directly above the German warships, but, before they could do anything about it, they were spotted by a number of fighters. Being totally outnumbered, all they could do was to aim for some cloud and then head for home to inform the authorities about the position of the German convoy.

A few minutes after the two 91 Squadron pilots had left the area, another two Spitfires arrived on the scene, flown by Grp Capt. Victor Beamish, who had recently been appointed as the CO of Kenley, and Wg Cdr Boyd. They too found themselves in the middle of the German fighter umbrella and were pursued by a large number of enemy fighters. These latter two Spitfires were also spotted by Beaumont and Oxspring, who were on their way home. Had they not seen the roundels on wings of the 91 Squadron Spitfires at the last moment, it could have easily resulted in a friendly fire incident.

When Oxspring and Beaummont landed back at Hawkinge at 10.50 a.m. and reported seeing a convoy, the true importance of what they had witnessed was not immediately realised. Only after Grp Capt. Beamish and Wg Cdr Boyd had returned was it finally established what they had seen and that the German capital ships were out. There was a plan that had been drafted to deal with such an event, but it was in a safe at Biggin Hill and the officer who had the key had gone off on leave. In the absence of any strategic organisation, even senior officers were not sure what to do and were forced to deal with events as they happened. Subsequently, there was very little co-ordination between services and departments and things went from bad to worse.

Manston's involvement began when Wg Cdr Constable-Roberts, the Air Liaison Officer on Admiral Ramsay's staff in Dover, rang Lt-Cdr Eugene Esmonde. According to the Manston ORB, the call was received at 10.55 a.m. Esmonde might have been half expecting such a call as he had volunteered the services of his unit for such an event, due to the experience that they had gained during attacks on the *Bismarck* in 1941.

After it had been confirmed that the Kriegsmarine's battleship convoy was cruising up the Channel, frantic messages were sent between 11 Group Fighter Command and Lt-Cdr Esmonde about which squadrons would provide fighter cover for his Swordfish. Just the day before, Esmonde had travelled up to London to receive his DSO from the King and he may well have pondered over how his circumstances had changed within twenty-four hours.

The six Swordfish of 825 Squadron were split into two flights, with 'A' Flight led by the CO and 'B' Flight led by Sub-Lt Thompson; they took off from Manston at 12.20 p.m., with Lt-Cdr Esmonde leading the formation in W5984 'H'. It is claimed that, as they got airborne, Wg Cdr Tom Gleave, the station commander, stood in the snow alone at the end of the runway and saluted each aircraft as it took off. This was a clear acknowledgement by the station commander of Manston that they were being sent on a 'suicide mission' and were not expected to return. Gleave and his staff had been heavily involved in the operation, passing on orders and communicating with all those involved.

With their 690-hp Bristol Pegasus engines, the Swordfish were barely capable of making 90 mph and were no match the Luftwaffe's fighters. It had been arranged that the Biggin Hill Spitfire Wing's 72, 124, and 401 Squadrons would provide close support to Esmonde's Swordfish and that they would rendezvous overhead Manston at 12.25 p.m.

Lt-Cdr Esmonde circled overhead Manston for several minutes after taking off, but only ten Spitfires of 72 Squadron, led by Sqn Ldr Brian Kingscombe, turned up. Knowing that the operation was time critical and not waiting for the rest of the escort to arrive, Esmonde led his formation out to sea towards the German convoy. According to the Manston ORB, the Swordfish were only 10 miles out when they were attacked by fifteen enemy aircraft, a mixture of Bf 109s and Fw 190s, and the 'odds were overwhelming'.

Although 72 Squadron pilots did their best to support Esmonde's Swordfish, they were actively engaged by the enemy fighters, mainly from JG26. The Spitfires of 124 and 401 Squadrons had been sent too far to the south, but eventually they turned up and joined the fray, while the two units of the Hornchurch Wing ended up over Calais.

The Manston ORB mentions that the air gunner of the 'second aircraft' was killed by machine-gun fire and that the observer had attempted to take over the gun, but could not get the gunner's body out of the way. What was described as the 'third aircraft' was hit by cannon fire and crashed onto the sea, but the crew were lucky to be picked up by a motor torpedo boat of the Royal Navy after one-and a-half hours in the sea.

Among the survivors from Swordfish W5983 'G' was Sub-Lt Brian W. Rose and Sub-Lt Edgar F. Lee, both of whom were later awarded the DSO. They were picked up by an MTB, No. 25, under the command of Lt Gamble, one of five MTBs that had left Dover at midday. The MTBs had already attacked the convoy with torpedoes and, although they were all badly damaged, they all managed to get back to port.

The other crew members that survived were Sub-Lt Charles M. Kingsmill, the pilot of Swordfish W5907 'L', and two members of his crew, his observer, Sub-Lt Reginald McCartney Samples and gunner, leading aircraftsman Donald A. Bunce. They were picked up by a minesweeper that had seen their aircraft crash into the sea. Kingsmill and Samples were awarded the DSO, while Bunce was awarded the CGM. Lt-Cdr Eugen Esmonde was awarded a posthumous Victoria Cross.

The observer of the 'third aircraft', 'Mac' Samples, later said that he had noticed that the fabric of his aircraft's wings had been torn by flak and was full of holes. Then a cannon shell hit the fuselage between Samples and Kingmill, wounding both airmen, while the gunner, LAC Bunce, was screaming insults at the Germans and shooting at an enemy aircraft—which he managed to shoot down. Feeling a burning sensation in his leg, Samples noticed a neat pattern of holes in his flying boots and, although blood was oozing out from his foot, he claimed that he had felt no pain.

As a consequence of his wounds, Samples had failed to notice that his pilot, Sub-Lt Kingsmill, had already dropped their torpedo, which had been aimed at the *Prinz Eugen* from a range of 2,000 yards. With the Swordfish on fire and badly damaged by flak, Kingsmill was struggling to maintain height and tried to communicate through the speaking tube, but it was shattered. Urgently needing to communicate with his pilot, Samples bravely climbed out of the cockpit and clambered up to the cockpit, then shouted to him while pointing down towards some MTBs of the Royal Navy. Over the noise of the engine and ensuing battle, Samples shouted to Kingsmill and told him that he should ditch by the boats. Probably realising that he had no other options, Kingsmill ditched on the water and minutes later the three men were pulled out of the sea. Samples was so badly wounded that he spent the rest of the war at RNAS Yeovilton, where he continued to receive treatment.

No. 217 Squadron also lost another two of its Beauforts and crews during Operation Fuller; they were among a formation of nine that landed at Manston at 2.53 p.m. to meet up with a number of Hudsons, which were being utilised to create a diversion while the Beauforts attacked. They took off at 3.34 p.m., but the Beauforts soon got separated from the Hudsons, which began their attack on the ships before the torpedoes were dropped. The Beauforts were unable to contact the Hudsons or the Spitfires because of a mix up on both W/T and R/T channels. The two Beauforts lost were AW278, flown by Flt Lt Finch, and L9877, flown by Flt Lt White, and at least another two Beauforts were lost from 22 and 86 Squadrons. Plt Off. Carson of 217 Squadron was awarded the DFC for his part in the operation.

Among other losses during Operation Fuller were those suffered by Bomber Command, and, with the exception of those in 5 Group, the squadrons had been stood down. Aircraft and crews had to be found and organised on the spur of the moment and Bomber Command flew 242 sorties to attack the German convoy with little hope of achieving any success. No. 142 Squadron, based at Waltham (Grimsby), sent six Wellingtons to attack the German convoy, but they did the sensible thing and turned back when it became known that the Luftwaffe had put up an umbrella over the ships. They were among the 188 aircraft from Bomber Command—mainly Hampdens and Wellingtons—that failed to make any contact at all with the German convoy.

Despite 142 Squadron's failure to find the battleships the following day, the Squadron received a signal from 1 Group HQ thanking its crews for their part in the operation. The same signal was sent to other units in 1 Group that had also failed to find the convoy:

Please convey to all who took part in yesterday's operation my warm appreciation of the efficiency and determination with which the attacks on the enemy warships were conceived and executed. I am sure that all units realise the supreme importance of keeping these German cruisers inactive and the great contribution their attacks have made towards relieving the Royal Navy of some part of its very heavy burden. It was most satisfactory that the number of enemy fighters destroyed by your gunners well exceeded your own losses.

Later that night, after reporting to Admiral Ramsey, Sub-Lt Lee was driven back to Manston where he was met by Wg Cdr Gleave, who shook his hand, but said very little about the operation. Due to his seniority and in the absence of Lt-Cdr Esmonde, Lee had assumed command and he had lots of things to clear up before he went off on leave. LAC Bunce also returned to Manston and visited the mess to collect some things that he had left there. One can only imagine what went through his mind as he walked through the buildings, where a few hours before things had been so different.

That evening, Gleave wrote up his report, in which he claimed that when he had spoken to Lt-Cdr Esmonde and his crews before they had taken off, they had 'displayed signs of great enthusiasm and keenness for the job they were about to undertake'. What else could he have said? The operation had not been a total disaster and both the *Scharnhorst* and *Gneisenau* had struck mines that had probably been laid by the Hampdens of Bomber Command.

A couple of weeks later, the *Gneisenau* was hit by two bombs that had been dropped by aircraft from Bomber Command while she was in a floating dock undergoing repairs—these repairs were never completed, leaving her a 'floating hull'. The *Scharnhorst* was repaired, but was never fully operational

The headstone of Lieutenant Commander Esmond, VC, in Gillingham Naval Cemetery. His body was washed ashore after he had been killed during the Channel Dash operation on 12 February 1942.

Air-sea rescue launch *127* that was based in Ramsgate and rescued many downed pilots.

again and was sunk by the Royal Navy at the end of 1943 by a number of British cruisers that included the *Duke of York* and *Belfast*. The Germans had effectively won the Channel Dash, but it had paid a high price, having both battlecruisers disabled for the rest of the war.

There are a number of memorials dedicated to the airmen of 825 Squadron and what became known as the Channel Dash. One of them is on the airfield at Manston behind the Spitfire and Hurricane Memorial Building, close to where the Swordfish took off. The main memorial is in Ramsgate Harbour, in front of the Maritime Museum, and it was organised by the Channel Dash Trust, who hold annual memorial services to ensure that those airmen and officers who took part are not forgotten.

7

Calm after the Storm

On 28 February 1942, Major Capen of the United States Army Air Corps visited Manston with a number of other officers to observe the procedure for attacks on shipping and particularly those tactics used by the Hurribomber squadrons. The title United Army Air Corps was about to change and be replaced by the force known as the United States Army Air Force. The American officers were given a lecture and shown a film of attacks on enemy shipping taken by gun cameras before they left later in the afternoon. This was the second visit by American Air Force staff since December, and it seems that they were willing to draw upon the RAF's experience and pass it on to its pilots.

A new section of WAAFs arrived at Manston on 3 March, but they were accommodated in the Ursuline Convent in Westgate because it was considered too dangerous for them to be based on the station. The officer in charge of them was Flt Off. Pam Barton, a professional golfer who had won the US Amateur Cup in 1936 and competed with the British team for Curtiss Cup in 1934 and 1936. She had also won the French International Ladies Competition in 1934 and had gone on to win the British Open Tournament in 1939.

On the outbreak of war, Pam Barton had volunteered as an ambulance driver and worked in London during the Blitz, but had later transferred to the WAAFs. She later trained as a radio operator and was awarded a commission.

In March 1942, 607 Squadron left RAF Manston with the unit destined for India via Margate Station and Southampton Docks. Equipped with the Hawker Hurricane IIa, the unit had arrived at Manston in October 1941 for experimental use of the aircraft in the fighter-bomber role. Its pilots had been enthusiastic about this commitment and, for his part, Commanding Officer Sqn Ldr M. J. Mowatt had been awarded the DSO.

Not all of 607 Squadron's personnel were destined to be posted overseas, and eight of its pilots, alongside a small number of ground staff, remained at Manston. Its seventeen Hurricanes, which had been used in the fighter-bomber role, were handed over to 174 Squadron, which was being formed at Manston, to use them in the same role as fighter-bombers with attacks on enemy shipping. The Commanding Officer of 174 Squadron was Sqn Ldr R. C. Wilkinson and the Squadron was allocated the code letters 'XP'.

Manston had a special visitor on 27 March, when Lt-Gen. Montgomery, who was then the GOC of 12 Corps, arrived to inspect the local defences. Montgomery had taken over the position on 27 April the previous year and he was no stranger to Thanet. Shortly after taking over, he had visited the area and had been surprised to find out that part of his command was an auxiliary unit (Churchill's Secret Army), which he had previously known nothing about.

On 2 April, just a few days after Montgomery's visit, a warning was received about a possible assault by German paratroopers on the airfield. The threat of invasion was taken seriously and, with 40 miles of weakly defended coastline around Thanet and east Kent, it was considered that it would have been ideal place for parachute troops or gliders to land.

The Army provided a battalion of troops to defend the airfield, which was considered to be a key objective if and when any invasion took place. Every able-bodied airman, whatever his trade—including cooks, drivers, and clerks—was forced to bear arms. Machine guns from wrecked aircraft were put on home-made mountings, along with a 30-mm cannon that was said to have once belonged to an experimental aircraft. The airfield was mined with Cortex that ran through conduits across the landing ground and other main areas where gliders might land. Thankfully, the invasion never took place and eventually Manston was soon able to return to its normal security state.

It was a sad day on 11 April, when Wg Cdr Gleave left Manston to attend the RAF Staff College and handed over to Wg Cdr Alfred Guy Adnams, who was posted in from the Air Ministry, where he had been a Director of Planning. It had been Wg Cdr Gleave's idea to have WAAFs posted to Manston to improve moral. He had also organised trips for the wives and families of airmen to be able to visit them in Thanet, especially those that were married. They had to obtain special clearance for wives and family members to enter the defence zone, but arrangements were made so that they could get passes for them also to stay in local boarding houses and hotels.

Wg Cdr Adnams had been awarded a Short Service Commission in 1928 and, after flying training at 5 FTS, he had served as a pilot in 19 Squadron, flying Armstrong Whitworth Siskins. In 1933, he had been given a Permanent Commission and soon afterwards served as a Qualified Flying Instructor and Flt Cdr in No. 4 Flying Training School. Several of his previous posts were related to administrative or technical work and his flying experience was quite limited compared to his predecessor.

RAF Manston took over the responsibility for another station in April, when the construction of a new GCI radar unit began a few miles down the road at Sandwich. The GCI radar was to replace the Chain Home system that only looked out to sea, and it could scan 360 degrees, looking inland to those areas missed by the older system. The first radar at Sandwich had been mobile equipment, but a decision had been made by the Air Ministry for Sandwich to become a permanent station. It was built on the south side of the Ash Road to the west of the town and it was not to be a self-accounting station, but operated as a lodger unit under the administrative control of RAF Manston.

In May, the size of the WAAF contingent was increased and made up for the fact that many of the airmen who had previously performed certain duties, such as clerical work, had been posted elsewhere. The women carried out a large number of different duties, but the main ones involved them being employed as MT drivers, fabric workers, telephonists, meteorological office assistants, clerical duties, and waitresses. Flt Off. Pamela Barton was eventually to have over 600 WAAFs under her command.

On 30 May, a contingent of 56 Squadron arrived on a detachment, with many of the ground crew flying in on a Handley Page Harrow. On the same day, four of the units Hawker Typhoons landed at Manston. Those four Typhoons were the first of their type ever to land at the airfield. It is understood that the previous station commander, Wg Cdr Tom Gleave, had been promised the Typhoons by the AOC of Fighter Command, Air Marshal Leigh Mallory, for the defence of Canterbury and East Kent.

The Typhoon had first flown in February 1940, but arguments about its development and problems with technical issues slowed down its entry into service. The Typhoon did not fly again until May 1941, and it entered service with the RAF in September that year with 56 Squadron. However, its entry into service was further delayed by technical problems and it only entered operational service on 30 May 1942, the day the unit arrived at Manston.

The following day, 56 Squadron flew its first patrols from Manston, with two Typhoons taking off in the morning and another two in the afternoon. Both patrols proved to be uneventful. On the same day, a Hurricane of 3 Squadron crashed into a hillside near Dover after taking off from Manston.

On 1 June, there was a serious incident involving friendly fire, when two of 56 Squadron's Typhoons were shot down by a section of Spitfires from 401 (Canadian) Squadron based at Gravesend. The two aircraft had been scrambled to intercept suspected intruders over Dungeness, as were two Spitfires of 401 Squadron, who arrived there first. The two Typhoons, R7678, with Plt Off. Deugo, and R8199, with Sgt Keith Mansell Stuart-Turner, had taken off from Manston at 6.20 a.m. Both the pilots of the 56 Squadron Typhoon and the 401 Squadron Spitfires were under the control of Biggin Hill.

The Typhoons were at between 25,000 feet and 28,000 feet and approaching Deal when the attack took place, but Plt Off. Deugo managed to bale out and spent two hours in the water before being rescued. Deugo, who was also Canadian, had sustained serious burns to his face and was badly wounded in the ankle where he had been hit by a cannon shell. Neither the body of Sgt Stuart-Turner or his aircraft were ever found. Ironically, one of the reasons that had delayed the Typhoon entering operational service was that there were fears its profile and appearance was very similar to that of the Fw 190. Such incidents proved that those fears were justified and, as a result, Typhoons had thick black and white bands painted around the wings in an attempt to identify them more clearly.

The next day, two more Typhoons of 56 Squadron carried out an uneventful coastal patrol. At the same time, six Hurricanes of 174 Squadron carried out Circus 181, attacking military objectives at the Forêt d'Eu. WO Merryweather's Hurricane was run into from behind by another aircraft and the fin, starboard tailplane, and top section of the rudder was swept away. Despite that, the WO bombed his objective and then returned safely to Manston, having flown 140 miles in what was essentially a wrecked machine.

The first ever Spitfire Mk VI landed at Manston at 5.30 p.m. on the 3rd, having flown on Circus 184. It was different from other Spitfires because it had a pressurised cockpit and pointed wing tips. The aircraft from 616 Squadron were used for high-altitude sorties to challenge enemy reconnaissance aircraft such as the Ju 86P-2 that could operate at heights in excess of 39,000 feet. The unit had been equipped with the type in April and although pilots spoke highly of its performance, they did complain that it was uncomfortably hot.

Manston had a royal visitor on 4 June; Air Commodore the Duke of Kent arrived in the early afternoon after having lunch at Doon House (St Peters), a former school in Westgate. A number of Spitfires from the Debden Wing landed during his visit, having returned from a Circus operation, and were in need of fuel. For the benefit of the Duke of Kent, the Typhoons of 56 Squadron carried out what was described in the ORB as an 'aerial line shoot'.

Two Beaufighters of 29 Squadron landed at Manston on the morning of the 6th after one of them had destroyed a Do 217 15 miles east of Sandwich. The pilot, Flt Lt Brian, and his observer, Plt Off. Gregory, had witnessed the aircraft crashing into the sea and exploding on impact.

A Handley Page Harrow arrived at Manston on the 13th to convey the ground crew and equipment of 174 Squadron that were leaving for a detachment to Ford. The Harrow left the following day with all the available aircraft and pilots taking off for the former Naval station at Ford in Hampshire.

Commodore Cunliffe, representing Vice-Admiral Dover, visited Manston during the afternoon of the 20th. At about the same time, three Hurricanes from 174 Squadron, which had only recently left for Ford, arrived back to be available for shipping strikes during the moon period. Two of them took-off at midnight, but one of them returned early because of R/T trouble; however, the operation had to be postponed anyway because of bad weather.

During the morning of 30 June, six Harrows arrived at Manston with the ground crews of the North Weald Wing; later in the day, the Spitfires of 222, 331 (Norwegian), and 332 (Norwegian) Squadrons arrived. No. 331 Squadron flew its first patrol on the 1 July, with two aircraft carrying out a search for shipping in the area of Ostend, but the pilots found nothing to report.

Despite the arrival of a number of different units, 174 Squadron's Hurricanes continued to operate between Manston and Ford and, on the 1 July, one of them was involved in a fatal incident. Shortly after 8 a.m., Sgt Smith took off in his Hurricane IIb to return to Ford, but he crashed into a hill near Dover. The accident was blamed on the weather and the fact that Sgt Smith had encountered low cloud and bad visibility.

Two more Harrows landed at Manston on 1 July, carrying the ground parties and equipment for 403 (Canadian) Squadron that was on a short detachment from its base at Catterick. The Harrows arrived at 1.40 p.m., but the nineteen Spitfires Vbs of the air party did not land until 5.10 p.m. because they had been delayed by bad weather.

The following morning, the pilots were down at their dispersal at 7.15 a.m. and were given a talk by the wing commander flying about local flying patterns and R/T procedures. They were also briefed about other restrictions that had to be observed as regards the strict security around the airfield. Bad weather prevented any flying taking place during the morning, but in the afternoon standing patrols were flown over convoy 'agent'. To the surprise of many on 403 Squadron during the early evening, they were off the camp and they were able to explore the local area and local 'watering holes', such as the Jolly Farmer and Prospect Inn.

During their short stay at Manston, the officers of 403 Squadron did their fair share of 'partying' and, on Friday the 3rd, there was a party in the Officers' Mess that went on from 10 p.m. until the early hours of the 4th. That was after everyone's spirits had been raised by an ENSA concert featuring Claude Herbert. Then, on Sunday night, Wg Cdr Adnams put on a 'farewell' party in mess, as he was about to hand over his role as Commanding Officer and Station Commander.

It is noted in the Manston ORB that, on 6 July, the majority of the Spitfires on the airfield had been camouflaged with two white stripes on the orders of Fighter Command. This was because they had been allocated to take part in the Dieppe Raid, known as Operation Rutter. On the same day, the AOC of Fighter Command, Air Vice-Marshal Leigh Mallory, visited the station to inspect the detachment from 3 Squadron. He also personally congratulated Fg Off. Hay and New Zealand Sgt Gawith, who had been awarded the DFC and DFM respectively.

The following day, 222 and the two Norwegian units, 331 and 332 Squadrons, returned to their home bases, while the order regarding the white stripes on the aircraft was reversed and ground crews were told to remove them immediately. There seemed to be some confusion as to where 174 Squadron was going to be based and, on the 7th, another Harrow arrived at Manston, in connection with the departure of 174 Squadron. On the 8th, two more Harrows arrived 'for the removal of 174 Squadron'. It seems that the unit spent several days at Fowlmere in anticipation of a posting overseas, which was then cancelled and the squadron returned to Manston. No. 1450 Flight was formed out of a cadre of 174 Squadron and began operations from Manston.

It was also on 7 July that 403 (Canadian) Squadron was informed that it would be returning to Catterick the next day. The ground party left Manston at 6.15 a.m. for the long drive back to Catterick and, because of various breakdowns, it did not arrive there until 10.45 p.m.

Wg Cdr Tom Gleave returned from his course at the RAF College and on 9 July took over command again from Wg Cdr Adnams, who went to take command of RAF Northolt. Gleave was the only officer ever to command RAF Manston on two occasions.

There was no operational flying on 28 July, but, during the evening, six Mosquitos of 23 Squadron arrived on a detachment. They would later operate against German-held Dutch airfields and as cover on a large raid on Hamburg by Bomber Command. Two additional Hurricanes arrived to join 3 Squadron and, later that night, five of its aircraft went out on Intruder sorties, from which two of its pilots failed to return. They were FS Shirm and New Zealander Sgt Gawith, who had only recently been awarded the DFM.

At 4.15 a.m. on 30 July, the first ever Avro Lancaster landed at Manston, its crew having dropped a 4,000-lb cookie and containers of incendiaries on objectives at Saarbrücken. The Lancaster belonged to 207 Squadron, based at Bottisford in Lincolnshire, and it had been one of the first units to receive the type.

Just five minutes after the Lancaster had landed, a Wellington from 460 (Australian) Squadron landed after being attacked by a Bf 109 during a bombing raid, killing the crew's rear gunner. At 4.40 a.m., a Stirling from 15 Squadron landed with engine trouble, it also having taken part in the raid at Saarbrücken. Once again, Manston was proving that it was an ideal location for an emergency landing ground. To co-ordinate the movements and diversion of its aircraft during bad weather, or those damaged by enemy action, a Central Flying Control had been established by Bomber Command. That meant that it was not always the pilot or crew who chose to land at airfields such as Manston, but sometimes the decisions of officers working in Flying Control at High Wycombe.

On 1 August, eight Hurricanes of 174 Squadron left Manston for Ford again, where five of them loaded up with 500-lb bombs and the remaining three loaded up with 250-lb bombs. At 12.20 a.m., they took off from Ford to attack a hutted camp in Normandy and they were escorted by the Hornchurch Wing. Unfortunately, the leader's compass was faulty and, being unable to find their objective, they had to be satisfied with attacking a goods train and the railway line.

An advance party from 23 Squadron arrived on 5 August, with the remainder of the unit flying in the next day. The Squadron had regularly operated from Manston with its Havocs and Bostons, but in July it had been re-equipped with the de Havilland Mosquito II. It flew its first sortie from Manston the next day, when Plt Offs Welch and Shuttleworth flew sorties to Gilze and Eindhoven, although the patrols proved to be uneventful.

At 9.03 p.m. on 9 August, the pilot of a 174 Squadron Hurricane spotted a Ju 88 over Margate, flying low at approximately 1,500 feet, although the Manston ORB stated that there were no enemy aircraft in the neighbourhood. It is now known that Margate was used as a datum point for the Luftwaffe's attacks on London, flying low across the Channel before setting course for the capital.

There were reports of a Short Stirling having crashed into the sea north of Margate during the early hours of 13 August, but then another Stirling,

Stirling BF325 of 149 Squadron that crash-landed in the back garden of a house in Rumfield Road and came to rest just feet away from what might have been a disaster. No one was seriously injured, and the aircraft provided a great deal of entertainment for local children.

BF325 'A' of 149 Squadron, entered the circuit with undercarriage problems. When the wheels failed to lower, the pilot, Sqn Ldr G. A. Watt, was ordered to make a belly-landing; however, when he found out that the undercarriage was down, he decided to make a normal landing. Unfortunately, as he was flying over Broadstairs, the engines began to fail because the fuel had run out and, at 3.55 a.m., the aircraft crashed into an Anderson shelter in the back garden of 132 Rumfield Road.

The aircraft came to a halt just feet from the back wall of the house—had it travelled any further, several properties would have been demolished and a number of residents killed. Several of the Stirling's crew were injured in the crash, but not seriously and they were shocked when they were told that there had been children playing in the shelter, where two of them were trapped there. However, none of them were seriously injured and a vigorous argument then broke out among them about what type of aircraft it was and if it was a Lancaster, a Halifax, or a Stirling.

8

The Dieppe Raid

The North Weald Wing arrived back at Manston on 14 August with a number of Harrow transport aircraft carrying the ground crews and support equipment. Later that afternoon, 331 (Norwegian) Squadron, commanded by Maj. Olrik Helge Mehre, and 332 (Norwegian), commanded by Maj. W. Mohr, Squadrons arrived along with 242 Squadron under the command of Sqn Ldr T. C. Parker.

Born in Narvik in February 1911, Mehre had graduated from the Norwegian Army Flying School in 1934, where he later became an instructor. When Germany invaded Norway in April 1940, he had escaped via Stockholm, Moscow, and the Pacific to eventually arrive at the Norwegian Air Force Training Centre in Toronto, known as 'Little Norway'. Having served on 242 Squadron, he was transferred to 331 Squadron, of which he had only recently taken command.

The Wing was under the overall control of Wg Cdr Scott-Malden, who had risen through the ranks quite rapidly from being a sergeant-pilot in June 1939 to being commissioned and achieving the rank of squadron leader by September 1941. It was Scott-Malden who had been given the job of helping to bring the first free Norwegian fighter squadron to operational readiness during November 1941 in Scotland. With the training having been completed, the Norwegians had moved south to become part of the North Weald Fighter Wing.

After the North Weald Wing had arrived at Manston, the Hurricanes of 174 Squadron flew to Ford, and, although technically still based at Warmwell, two of them would return to Manston the next day to be at readiness. The unit continued to regularly use Manston as a forward air base.

At 2 p.m. on the 15th, 403 (Wolf) Squadron arrived from Catterick again, and the unit, equipped with the Spitfire VB, had to travel through North Weald because of the weather. Two aircraft overshot the runway while trying to land and, while Plt Off. Johnson was lucky to escape being injured, Plt Off. Anderson broke a shoulder blade. Formed in September 1941, the unit was the first to be formed overseas and it had no connections with any other

Canadian squadron. Up to the 16th, the Squadron had been commanded by a New Zealander, Sqn Ldr Al Deere, but he had been given a staff job and the new CO was Sqn Ldr Leslie Sydney Ford.

Ford had joined the Royal Canadian Air Force in June 1940 after attending Acadia University in Nova Scotia, and he had been posted to the UK in February 1941. After training at 52 Operational Training Unit, he had been posted to 403 Squadron, which had just been formed before being posted to 402 Squadron when it was equipped with Hurricanes in the fighter-bomber role. He had also served on 175 and 412 Squadrons and, in June 1942, was awarded the DFC, taking command of 403 Squadron just a few days before the Dieppe Raid.

The following day, the second ground party arrived from Catterick and 403 Squadron flew its first operation under the leadership of Sqn Ldr Ford; this first operation was a Circus in the area of St Omer at 12.45 p.m. Having returned safely at 1.40 p.m., the pilots were immediately briefed for another Circus operation and they took off at 4.35 p.m. to operate over Dunkirk. No enemy aircraft were encountered and the Squadron landed at 6.05 p.m. with Sqn Ldr Ford having nothing to report.

On the 18th, the final preparations for the Dieppe Raid were being made at Manston and, during the afternoon, Grp Capt. Morris, the station commander of North Weald, and Wg Cdr Scott-Malden arrived. They assembled all the pilots of the squadrons that were going to take part in the mission in the station's Intelligence Officer's briefing—Operation Jubilee was going to take place the next day.

The operation involved 5,000 troops of the 2nd Canadian Division that were under the command of Maj.-Gen. Hamilton Roberts to assault Dieppe by first raiding German positions to the east and west of the town. The reason the Canadians had been chosen was that they had been in Britain since the end of the 1939 and, after endless training exercises, many of them were getting restless and wanted to get into action. Some 230 ships that sailed mainly from Southampton and Newhaven were to take part in support of the operation.

The RAF's contribution, including those units from the supporting Commonwealth countries, was seventy-four squadrons, sixty of them being fighter units. Many of the aircraft were Blenheims or Bostons from 2 Group Bomber Command. The Dieppe Raid was mainly a Canadian operation on the ground, although they had a heavy presence in the air too. Canada was also represented by a total of eight air force units.

Later on during the evening of 18 August, eight P-51 North American Mustangs that were to take part in the operation arrived at Manston, two each from 26, 239, 400, and 414 Squadrons. These aircraft were the Mustang Is and were fitted with the Allision engine that had a poor performance at altitude, especially compared to those aircraft that were later fitted with Merlin engines. There was no operational flying on the night before the operation.

One of the main aims of the Dieppe Raid was to test the German defences, but also for the RAF to provide air cover for the Royal Navy and the Army. Fighters and fighter-bombers were ordered to attack the coastal defences and to lure aircraft of the Luftwaffe away from the action on the beaches.

At Manston, the morning of 19 August was fine, with just a slight mist across the airfield. The Spitfire squadrons that were to take part in the operation—242, 331, and 332 Squadrons—were at readiness from 5 a.m. Between 5.30 a.m. and 6.00 a.m., the Mustangs took off, and at 6.15 a.m. thirty-six Spitfire VBs set out for Dieppe, led by Wg Cdr Scott-Malden. Nos 331 and 242 Squadrons carried out patrols over the Channel at 5,000 feet, while 332 Squadron stayed just below cloud at 13,000 feet, but later had to descend to help 331 Squadron, which was soon engaged by the enemy.

All three units were heavily engaged by numerous Fw 190s, although both Norwegian units claimed a number of enemy aircraft destroyed. Both Maj. Mohre and Capt. Birksted of 331 Squadron claimed Fw 190s destroyed and the unit's total tally was six Fw 190s. No. 332 Squadron claimed to have destroyed five Fw 190s, with another three claimed as probables or damaged.

Two of 332 Squadron's pilots, Sgts Bergeland, in Spitfire AB629, and Staubo, in BL819, failed to return to Manston. It was later found that both men had become POWs. Sgt Raeder of 332 Squadron was wounded in the foot and had to make a forced-landing in Sussex. Second Lieutenant Greiner was forced to bale out over Dieppe, but managed to evade capture and returned later in the evening courtesy of the Royal Navy.

The Germans made determined efforts to attack the shipping involved in the operation and it was noted that a number of Do 217s were involved in the bombing raids, many of which were unescorted. The Norwegian units claimed to have decimated these, destroying six of them, as well as damaging another.

The twelve Spitfires of 403 Squadron took off from Manston at 6.47 a.m. to provide top cover over the town of Dieppe between 7.20 a.m. and 7.50 a.m. No. 403 Squadron missed out on what was considered to be the easy prey of the unescorted Dornier bombers because it was fiercely engaged by the fighters, a mixture of Fw 190s and Bf 109s. It lost three aircraft and Plt Offs Gardiner, Walker, and Monchier failed to return. In return, 403 Squadron pilots claimed a single Fw 190 and a Bf 109 destroyed.

A number of pilots were wounded but made it back to Manston, and Norwegian Maj. Moehre was hit in the leg and Sgt Walker of 242 Squadron was also wounded. Walker overshot the runway at Manston and crashed into a hangar that was only half built, but he still survived. At 11.15 a.m., thirty-five Spitfires from 331, 332, and 403 Squadrons took off to patrol Dieppe between 5,000 and 7,000 feet and, rather remarkably, thirty-four of them returned safely and landed at 1.15 p.m. The only pilot to be lost was Sgt Loftaggard, who was forced to bale out, but survived and later returned to England.

At 12.15 p.m., twelve Spitfires from 242 Squadron took off and returned at 2.10 p.m. after what was described as an uneventful patrol; at 2.15 pm., the two Norwegian squadrons took off again and patrolled over the convoy from 2.15 p.m. until 3.30 p.m. During this time, it was noted that the ships and support vessels appeared to be on their way back to England. Lt Berg and Sgt Djonne of 331 Squadron were forced to bale out, but both airmen were lucky and managed to get back to British shores by boat. During these sorties, 331 Squadron claimed three Fe 190s Destroyed and three damaged, while 332 Squadron claimed two damaged. There was also another 'friendly fire' incident involving a Typhoon from 266 Squadron that was shot down by a Spitfire of 332 Squadron.

The Typhoon, serial number R7815, flown by twenty-six-year-old Plt Off. Roland Herbert Leslie Dawson from Salisbury, Rhodesia, and he was flying with 266 Squadron as part of the Duxford Wing. Rather ironically, Dawson was the only pilot from the Duxford Wing who had had any success during the Dieppe operation and, shortly before the 'friendly fire' incident, he had shot down a Do 17.

As a result of this incident, Typhoons were painted with yellow wing bands to make identification easier and for them to stand out more clearly from the Fw 190s, with its similar profile. The yellow bands were later replaced with black and white bands, but, despite that, Allied pilots were often confused and the Typhoon was plagued by the problem of identification. No. 266 Squadron lost two Typhoons during the Dieppe Raid, the other being R7813, flown by Plt Off. W. S. Smithyson, who was also killed.

Another twenty-four Spitfires, twelve each from 242 and 403 Squadrons, took off from Manston and patrolled over the convoy between 4.50 p.m. and 5.40 p.m. However, they had all landed by 5.50 p.m., with 242 Squadron having nothing to report. No. 403 Squadron was engaged by a number of Fw 190s, with Sqn Ldr Ford and Plt Off. Murphy each claiming one destroyed, while Sgt Fletcher was credited with damaging another.

At 6.50 p.m., twenty-four Spitfires from 331 and 332 Squadrons, led by the leader of the North Weald Wing, Wg Cdr Scott-Malden, took off to patrol the final stages of the convoy's return, which was by then close to British shores. There were very few enemy aircraft around, but Lt Sem and Sgt Fossum of 331 Squadron spotted and chased and a single Do 217, which they claimed to have damaged. All aircraft had landed safely by 8.20 p.m.

The final sorties of the day from Manston were another joint effort by 242 and 403 Squadrons, which contributed twelve Spitfires each and carried out an uneventful patrol over Beachy Head between 7.25 p.m. and 8.30 p.m. Manston's contribution to the Dieppe Raid was twenty enemy aircraft destroyed, with another four probables, and sixteen damaged for the loss of four pilots, nine Spitfires, and three wounded.

That was not the full story, however, and, although historians have argued over the exact figure, it is generally accepted that the RAF lost a total of 106 aircraft, eighty-eight of them being fighters. There are the serial numbers of 109 Allied aircraft that were claimed lost on 19 August, among them a small number of Bostons and Blemheims from 2 Group.

Of those units that suffered the heaviest losses were 3 Squadron, which lost four Hurricanes and two pilots, and 26 Squadron, whose Mustangs were operating on reconnaissance sorties and lost all five of its aircraft. Three of its pilots were killed, while the other two were taken as POWs. No. 174 Squadron, who had strong connections with Manston, also lost five Hurricanes, with just two pilots surviving, one of whom became a POW. Apart from the loss of aircraft, the operation took its toll of experienced pilots such as Sqn Ldr A.E. Berry, the CO of 3 Squadron, and Sqn Ldr E. F. L. Fayolle of 174 Squadron. It was far more difficult to replace an experienced pilot than it was an aircraft such as a Hurricane or Spitfire.

There were lots of lessons learned from the failed Dieppe Raid, not least because the beach and ground where the attack took place was not suitable for heavy armoured vehicles such as the Churchill tank. After offloading onto the shingle beach, it was found to be virtually impossible for the tracks of the tanks to get a grip and many were abandoned without seeing any action. Investigation and reconnaissance of those beaches later selected for the D-Day landings were a vital part of its success, and, many months prior to the operation, samples of sand were brought back and tested to see if it was suitable to bear the weight of armoured vehicles.

Recent research has discovered that the Dieppe Raid was not just about capturing the port and knocking out the guns so that HMS *Locust* could enter the harbour, but to demonstrate to the Germans what a commando raid could achieve. It is claimed that it was actually a 'pinch raid' and the true purpose of the operation was to capture German code books and documents relating to the Enigma machines, if not an Enigma machine itself. One of the units involved was a special intelligence team led by James Bond author Ian Flemming, and its role was to guide troops to certain buildings, such as German Naval HQ, and capture documents and equipment held there. A lot of Canadians died thinking that they were just taking part in an elaborate commando raid, but the truth was a lot more complicated—and it was kept hidden for many years.

The RAF learned a lot of valuable lessons from the Dieppe Raid that would be put into practise for D-Day in 1944, and one them was the need for forward air controllers to direct the fighter-bombers to specific targets. Also the use of the 'cab rank' system, where the fighter-bombers, such as the Typhoon and Tempest, would be called in as required, while others orbited the target waiting to attack.

The Typhoons and Tempests were much more suitable types of aircraft in the ground-attack role than the Spitfires that were used on the Dieppe Raid. Operation Jubilee may have gone down in history as a total failure, but if it had not taken place, the mistakes made in August 1942 may well have been made in June 1944, making Operation Overlord and the invasion of Europe a disaster.

9

Any Port in a Storm

August 1942 was an extremely busy month, with a higher number of crashes, emergencies, and unexpected arrivals than normal. No. 23 Squadron were involved in a couple incidents and, on the 23rd, Sqn Ldr Starr crashed while trying to land his Mosquito after losing an engine. Struggling to control the aircraft, he overshot the runway before crashing into a building occupied by the Bofor gun crew. Fortunately, nobody was seriously injured.

The following day, there was another first for Manston, when a B-17 Flying Fortress from USAAF's 340th Bombardment Squadron, 97th Bomb Group, 1st Bombardment Wing, landed at the airfield. The aircraft, 41-9175, had been badly shot up while taking part in Circus 208, an attack on the dockyards at Le Trait. The Fortress was a B-17E, a modified version of the original design that incorporated a wider rear fuselage and a rear turret. Three of the crew were injured and this was described in the ORB as 'not only the first Fortress to land at Manston, but the first American operated aircraft'. This was just one week after twelve B-17s of the 97th Bomb Group had carried out the first heavy bomber raid on marshalling yards at Rouen, France, led by Maj. Paul Tibbet.

The night of 28 August was a turning point at RAF Manston, in so far as that what happened tested its facilities and resources to the hilt. The night began when a twin-engined Armstrong Whitworth Whitley bomber crashed close to the airfield. Fortunately, the crew were not injured, but they were checked out into the sick quarters anyway. At 8.15 p.m., without any prior warning, the Northolt Polish Fighter Wing began to arrive, which comprised of Spitfires from 302, 306, 308, and 317 Squadrons; by 8.55 p.m., forty-five of them had landed. Added to the action and unannounced arrivals were a large number of movements by 3 Squadron's Hurricanes and 23 Squadron's Mosquitos.

Just before midnight, a Wellington from 305 Squadron crash-landed on the flare path. It had set out to bomb objectives at Saarbrücken, but it had been attacked by three night-fighters before it had reached the target. The pilot had jettisoned the bombs after the rear gunner had been killed. The navigator, wireless operator, and front gunner had baled out over enemy territory and the aircraft had been flown back by just the captain and second pilot.

At 12.30 a.m., a Wellington from 101 Squadron landed and, having failed to reach its target in Nuremberg, the aircraft was desperately short of fuel. Another Wellington from 460 Squadron landed at 1.50 a.m. and it had also run out of fuel. The pilot overshot the runway at Manston and the aircraft was totally wrecked, but the crew escaped without serious injury. Another Wellington from 115 Squadron landed at 2.30 a.m. with another crew that had failed to find their objective at Saarbrücken.

It was then the turn of the larger four-engined Stirlings, with the first one from 149 Squadron landing at 3.30 a.m. and another one from 214 Squadron at 3.44 a.m.—this had one unserviceable engine and another failing. At 3.57 a.m., another Stirling from 15 Squadron landed and, at 4.30 a.m., another, N3717, from 218 Squadron landed. The pilot of this last Stirling had been told specifically to avoid the obstructions caused by the Wellington that had crashed a short time before.

However, the 218 Squadron pilot, Plt Off. Du Toit, claimed that he had been dazzled by the chance lights and landed too far to the right of the grass runway, sweeping through a line of Spitfires from the Polish Wing, completely writing off at least one of them (which burst into flames) and damaging many others. Despite the efforts of the station's fire service and local fire fighters that had been called in, very little could be done to save some of the Spitfires.

At the same time as all that was going on, another Stirling from 7 Squadron, R9158, flown by Sgt R. H. Middleton, landed and it was so low on fuel that its pilot was trying to land before the engines cut out completely. Like the previous Stirling, it landed too far to the right of the runway and, although the pilot managed to avoid the Spitfires, he could not avoid a large wooden hut. The Stirling crashed through the hut and its port wing then struck the corner of a Bellman hangar, finishing up against the wall of the station armoury. The aircraft was a Cat 'E' write-off, and the accident was later attributed to engine failure and a shortage of fuel.

Fortunately, the two airmen who normally occupied the hut were out working on the airfield; had it not been for the chaos caused by the arrival of so many aircraft, they may have been killed. Many other airmen had lucky escapes and a flight sergeant, who was co-ordinating things on the airfield, claimed that he nearly had his head taken off by the wing of the Stirling and the medical officer (MO) had to run for his life to get out of the way the aircraft. At 4.15 a.m., another Wellington from 101 Squadron landed and at 4.30 a.m. another Stirling from 218 Squadron also landed, but thankfully that was last aircraft to land that night. This final Stirling to land was so low on fuel that its engines stopped as it touched down because it tanks had run dry and it had to be towed off the runway.

The following morning dawned fine and between 6.45 a.m. and 7.15 a.m. what was left of the Polish Fighter Wing took off to return to Northolt. The north-western corner of the airfield was littered with wrecked Wellingtons and Stirlings and many others, including serviceable aircraft. Many of them had arrived from the parent airfields of those crews that had landed during the night, to collect them and take them back to their respective bases. The airfield and many of its hangars and buildings were in a terrible state of disrepair and, early the next morning, Wg Cdr Gleave walked among the chaos that still prevailed, ordering that photos were to be taken.

Among the debris was a Fairey Albacore of 841 (FAA) Squadron that had only recently arrived at Manston from Middle Wallop; the Naval biplane had been crushed beneath a Stirling, along with a Spitfire that had been flattened by a wheel from a Stirling. Photos of the disastrous scene were taken from every angle and a full report, compiled by Wg Cdr Gleave, were sent to the AOC of Fighter Command, AVM Leigh-Mallory. The report was then passed on to the Air Ministry for immediate action.

No. 174 Squadron was released from operational duty at Manston from noon on 30 August because it was to move out to Warmwell in Dorset the following day. The unit had been closely associated with Manston since March, when it had been formed at the station with Hurricanes; although it was posted out, it would regularly use the airfield as its forward base. The following spring, it would be re-equipped with the Typhoon and would continue to play an important part in the air war until it was finally disbanded in March 1946.

With the failed Channel Dash operation still fresh in the minds of many of those at Manston, four Swordfish of 819 Squadron (FAA) arrived during the early hours of 5 September. They had carried out an operation against a ship of approximately 5,000 tons that had been attacked by eleven other vessels. Another six Swordfish from the same unit landed later in the day and for those who saw them it must have brought back evocative memories of that fateful day in August when so many airmen had been lost.

Wg Cdr Tom Gleave, one of those who had been closely involved in the events of that day, handed over command of the station to Wg Cdr Charles Ronald Hancock, CBE, DFC, on 6 September. Having served at the station on two occasions, he is the CO who is best associated with RAF Manston and the one who most made his mark. Before leaving Manston and especially in light of the events of the night of 28–29 August, Gleave pleaded with the Air Ministry to build an asphalt or concrete runway, long enough and wide enough to accommodate those aircraft of Bomber Command that were damaged and in need of a safe haven.

On leaving Manston, Wg Cdr Gleave was posted to the planning staff of what later became called Operation Overlord, the invasion of Europe, and soon afterwards he was promoted to group captain. From October 1944 until

July 1945, Gleave went on to serve as General Eisenhower's Head of Air Plans and later Senior Air Staff Officer to the RAF's delegation in France. In 1953, he was invalided out of the RAF after undergoing more plastic surgery at East Grinstead Hospital. Having already been elected a Fellow of the Royal Historical Society, he became Deputy Chairman of the Battle of Britain Fighter Association. Grp Capt. Tom Gleave died in June 1993, aged eighty-four.

Wg Cdr Hancock (05112) was an experienced officer, who, as a young pilot officer, had been elected to the Royal Aero Club in August 1925. He had been awarded the DFC for action on the north-west frontier in 1930. His previous command was that of 16 Squadron from 1940 to 1941, when it had been equipped with Westland Lysanders. Hancock had been promoted to the rank of squadron leader in 1936 and wing commander in January 1941, being awarded the CBE at the same time.

It was on the 6th that a B-17 Flying Fortess, 41-9026 *Baby Doll*, became the first of its type to ditch into the English Channel—all nine crewmen were killed. The aircraft was from the 342nd Bomb Squadron, part of the 97th Bomb Group.

Aircraft from Bomber Command continued to arrive at Manston, and, on 9 September, a Wellington of 150 Squadron landed at 2.00 a.m. short of fuel. At 2.50 a.m., the ominous sound of grinding metal was heard across the airfield as two Stirlings landed at the same time—the first one from 214 Squadron collided with the other one from 7 Squadron. As the 214 Squadron Stirling landed, its undercarriage collapsed, one of its engines fell out of its mountings, and the aircraft was badly wrecked. Despite extensive damage, the only injury was to the pilot, Plt Off. Ince, who suffered a slight head wound.

The Westland Whirlwinds of 137 Squadron arrived from Snailwell in Cambridgeshire during the morning of 17 September; the unit had been posted to Manston under the command of Sqn Ldr H. St. Coghlan. The Whirlwinds were twin-engined fighters that had flown for the first time in May 1940 and entered squadron service with 263 Squadron in July that year. Despite its armament of four cannon, the aircraft did not have a very good reputation; during the Channel Dash operation, the squadron had lost four of them.

Not all those aircraft from Bomber Command that were in difficulties made it to the safety of Manston or even to the relatively safe haven of land—even someone's back garden. During the early hours of the 20th, at 4.30 a.m., a Stirling crashed into the sea off Margate. A Walrus from 77 Squadron based at Martlesham Heath later landed at Manston after searching for any survivors. The crew of the Walrus had managed to rescue three airmen from the wreckage before it sank beneath the waves. The body of another crew member was found later, but the search and rescue team from the Walrus failed to locate the bodies of the other two airmen, who were both known to have climbed out of the Stirling.

A Westland Whirlwind twin-engined fighter of 137 Squadron at Manston.

Broadstairs came under attack on 28 September by an unidentified aircraft of the Luftwaffe that was hidden in thick cloud. The following day, a Beaufighter from 29 Squadron was sent to Manston just in case there should be a repeat of the previous day's events. The Luftwaffe failed to turn up, however, and the Beaufighter returned to West Malling later that evening.

Nos 331 and 332 Squadrons continued to operate from Manston in October and, on the 6th, three Spitfires from 331 carried out a search for a torpedo boat that had been reported missing, but nothing was found. During the morning of the same day, 137 Squadron was called into action, but only to destroy a rogue Naval balloon that had been drifting over Birchington.

On 9 October, another B-17, 41-24362, made the first successful ditching of a B-17 in the European theatre of operations when it came down in the sea off the coast of North Foreland. The aircraft, captained by 1Lt Dinald M. Swenson from the 419th Bomb Squadron, 301st Bomb Group, had bombed an industrial target near Lille with an illegal crew member on board. Master Sergeant Glenn Doerr was a member of the ground crew who had persuaded Swenson to take him on an operation without obtaining official permission. However, Doerr was so influential in aiding other members of the crew to escape from the aircraft that he avoided being charged with a breach of regulations.

No. 137 Squadron, along with 174 Squadron, was called upon again on 9 October, but this time in an operational role in anticipation of E-boat activity in the Channel. At 7 a.m., the three Norwegian squadrons, 330, 331, and 332, were briefed to take part in Circus 224, which was to be led by Wg Cdr Smith, who was flying with 331 Squadron. The wing took off at 7.45 a.m. and climbed to 15,000 feet to rendezvous with two other wings and a squadron of Defiants over Felixstowe at 8.15 a.m. The Norwegians returned to Manston at 9.05 a.m., having suffered no casualties and making no claims.

On the same day, a squadron of P-38 Lockheed Lightnings arrived over the airfield to rendezvous with 108 Liberators and B-17s that were taking part in a Circus operation against Fives-Lille's engineering works. The P-38 twin-engined, twin-boom Lightnings did not land, but flew low overhead; this was the first time they had been seen at Manston. The formation returned at 10.30 a.m. after the eleven Spitfires that had escorted the bombers from a number of different units landed at Manston.

During the night of 24–25 October, Manston was invaded by aircraft from Bomber Command again. Night activities began at 1.30 a.m. and aircraft continued to arrive until 3.48 a.m. There were four Stirlings, two from 15 Squadron and one each from 149 and 218 Squadron. Additionally, two Halifaxes from 10 Squadron, based at Melbourne, Yorkshire, and another Halifax from 405 Squadron, based at Topcliffe, arrived at the airfield. A single Wellington from 75 Squadron at Mildenhall was the final unexpected visitor of the night. Apart from a Halifax from 10 Squadron, which had burst a tyre on landing, and one of the Stirlings from 15 Squadron, which had undergone an engine fire, all the aircraft left the next morning.

That night, seven Wellingtons of 142 Squadron arrived at Manston, plus another one that carried ground crew and technicians. One of the Wellingtons had bent its tail wheel fork while landing, but the others took off that night on a bombing operation to Milan. They were part of a force of twenty-three Wellingtons, twenty-five Halifax, and twenty-three Stirlings that took part in operations that night. At this point, Manston was being used by Bomber Command to directly despatch its aircraft on operational sorties—a new role for the station. It was almost certainly being done to save fuel as it was some 200 miles closer to Europe than other airfields in Yorkshire and Lincolnshire.

The Whirlwinds of 137 Squadron flew the unit's first offensive bomber patrol from Manston on 31 October, but, of the four aircraft despatched, only a single machine returned. A Walrus from 277 Squadron was sent to search the area, but only one of the three missing pilots was found, 5 miles off the French coast in the area of Le Toquet and Boulogne.

After the victory at El Alemein (Second Battle of El Alemin) by Lt-Gen. Montgomery's Eighth Army in November 1942, there was some reason for optimism among the military, especially the Desert Air Force that had played a crucial part in Montgomery's campaign. The disasters of the Channel Dash and the failure of the Dieppe Raid had not been forgotten, but lessons had been learned. There were many changes that had improved the RAF's operational capability, including the introduction of new types of aircraft such as the Hawker Typhoon.

The Advance Guard of 609 (West Riding) Squadron arrived at Manston on detachment from 1 November with its Typhoons, under the command of Sqn Ldr Roland Beamont. In December 1941, the squadron leader had been attached to Hawker at its Langley factory, where he had flown and worked on the Typhoon as a test pilot. It is widely recognised that he was instrumental in solving the many problems associated with the aircraft's engine and airframe that delayed the Typhoon's entry into service.

Although many of the ground crew were from Yorkshire, 609 Squadron was effectively an Anglo-Belgium unit. Flt Lt Jean Michel de Selys Longchamps was one of 609 Squadron's pilots; he was also a baron and the second cousin of the King of Belgium. He had first served as a Belgian cavalry officer and he escaped from France with the BEF at Dunkirk, but made the mistake of returning there before the country finally capitulated. Baron de Selys Longchamps had been captured by the Vichy and imprisoned at Marseilles, but he managed to escape and make his way back to England, where he was given flying training and ended up joining 609 Squadron.

What began as a short detachment for 609 Squadron at Manston grew in to a 'posting' and the unit and its CO would go on to be closely associated with RAF Manston. The remaining Typhoons of the unit followed on to the airfield the next day, with the aircraft flying its first sorties (coastal patrols) on the 3rd, but its main role would be as a low-level interceptor.

Wg Cdr Hancock handed over command of the station to Wg Cdr Walter Charles Curtiss Sheen on 10 November, after Hancock had been injured in a car crash. The thirty-five-year-old wing commander (Sheen) had joined the RAF in 1923 as the 7th Entry of aircraft apprentices at No. 1 School of Technical Training, Halton. He had gone on to train as a pilot and served as a sergeant pilot before being commissioned in 1930. He had later become a QFI and, in 1934, a member of the CFS's aerobatic team; he later went on to command 106 Squadron, which was his position when the war broke out.

A Wellington of 150 Squadron, serial number BK538, based at Kirmington, crashed in flames on the night of 20 November, killing four of the five airmen on board. The aircraft was returning from a sortie to Turin and, at 11 p.m., the pilot, Plt Off. J. F. Sweet, had overshot the runway while trying to make an emergency landing. Rather remarkably, the rear gunner, Sgt E. O. Booth, survived and was found sitting in his turret unharmed.

At 2.30 p.m. on 12 December, a Spitfire from 340 (Free French) Squadron landed at Manston after the aircraft had ran out of petrol while approaching Richborough. It had been in the air for two hours and forty minutes as part of an escort wing covering bombers that had been on operations over France—the pilot should have landed at Hawkinge. The French pilot had got lost and flown straight past Hawkinge, but, running short of fuel, he had spotted Manston and glided for over 10 miles to make a text-book landing without damaging his aircraft.

Three German sailors were rescued on 14 December from a dingy that had been spotted floating in the sea 10 miles south-east of Dover; they were spotted by a Typhoon pilot of 609 Squadron, who was on a coastal patrol from Manston. A Walrus was despatched and it was given air cover by two Spitfires from 91 Squadron, while the air-sea rescue crew arrived on the scene. The Walrus crew found that there were six men in a life raft and they struggled to get them all on board the aircraft. Unfortunately, because of very rough seas, the aircraft was repeatedly lifted up on to the raft, upsetting it and throwing the men back into the water.

The crew of the Walrus eventually succeeded in getting three men back on board, but could not take on the others because of the amount of water the Walrus had shipped. To add to the crew's difficulties, they found that, due to the rough sea, the aircraft could not get airborne and it took over an hour for it to taxi into Dover harbour. By the time that it had returned to collect the other three sailors, darkness had fallen and the Walrus crew were unable to find them. It was later discovered that the men were German sailors from a ship that had been sunk a few days before by a British destroyer.

No. 609 Squadron lost an aircraft and one of its pilots on 15 December, when Typhoon R7680 ditched 2 miles south-east of Ramsgate; Fg Off. Henry Desmond Fitzmaurice Amor was killed. The body of the twenty-one-year-old pilot from Hereford was washed ashore a number of days later and buried in Margate's St John's cemetery.

It was an important day on 22 December when Air Commodore Charles Curtiss Darley, a member of the Airfield Board at the Air Ministry, visited the station to assess and examine the feasibility for the provision of a runway 9,000 feet long and 750 feet wide. His report stated that it might be possible for the new east–west runway to be extended and rebuilt to meet modern requirements. This is what the former Commanding Officer, Wg Cdr Gleave, would have wanted, but as he was no longer at Manston it was left to Wg Cdr Sheen to deal with. He had also requested some changes to the airfield, but mainly just to extend the taxi tracks.

The work that Sheen wanted done had been estimated to cost £5,000, but Air Commodore Darley was not impressed and said that by diverting the main road and knocking down a number of houses, a new runway could be built. Darley told Sheen 'to think big', which would involve rebuilding virtually the whole airfield at a cost of approximately £1 million. Darley certainly knew how to get things done.

This was, in effect, Wg Cdr Gleave's legacy at RAF Manston, a runway that would be 9,000 feet long and 750 feet wide, one of only three in the country built to such specifications, with the other two being at Woodbridge, Suffolk, and Carnaby, Yorkshire.

It would be quite a while until the new runway was built. As a result, in order to help the pilots land in bad weather, a system known as FIDO (Fog

Intensive Dispersal Operation) was installed down each side of the existing runway. The intense heat from the fires fuelled by paraffin caused any fog or low cloud to disperse, break up, and allow the pilot a clear view of the runway.

Air Commodore Darley was a civilian, who had been discharged from the RAF in September 1939 after serving in the forces for nearly thirty years. He had attended the Royal Military Academy in 1909 and later served in the Royal Field Artillery in India, before returning to England and gaining his Royal Aero Club Certificate, No. 592, on 15 August 1913. He had flown in the First World War and, in October 1915, had survived being shot down in his Vickers F.B.5 'Gunbus' by no less a pilot than German ace Max Immelmann.

On 27 September 1919, the then Flt Lt Darley had been flying as navigator in a Vickers Vimy from England to Egypt with his brother, Capt. Cecil Hill Darley, as the pilot, when, for technical reasons, the aircraft made a forced-landing near Lake Bracciano in Italy. The following morning, they took off again, but Capt. Darley misjudged the take-off run and the aircraft hit a telegraph pole, crashed, and burst into flames. Flt Lt Darley ignored the immediate danger and tried to rescue his brother, who was pinned into the pilot's seat, but he was unable to get him out alive.

Darley had sustained such serious burns to himself that he spent eighteen months in hospital recovering from his injuries, but he was later awarded the Albert Medal for his bravery. In 1938, he had been appointed as the AOC of No. 1 (Indian) Group. However, after being involved in another air crash, had been invalided out of the RAF in September 1939 on medical grounds. With his vast amount of aircraft experience, Darley was given a post in the Air Ministry.

Manston was closed for a short while on 29 December after an incident involving a Fairey Albacore of 841 (FAA) Squadron—aircraft No. BF772. Lt Garthwait had just taken over command of the unit and he was on his first flight when the tyre on one of the wheels of the Albacore blew out on take-off, which caused the undercarriage to collapse. In turn, it set off one of the flares the Albacore was carrying, setting fire to the aircraft and forcing Lt Garthwait and Sub-Lt Waiting to run clear of the blazing wreck.

The two Naval officers feared that the petrol tank and the ordinance that the aircraft was carrying could explode at any moment—they were not wrong. Initially, the four 100-lb bombs exploded, quickly followed by the four 250-lb bombs; this created a scene of carnage all around them. The runway was effectively blocked and the airfield was closed temporarily, but during the evening a flare path was lit for the use of those aircraft those pilots or crews that had to land in an emergency.

The last day of 1942, Thursday 31 December, was the same as that of the day before, when the first snow of the year had fallen on the airfield. No. 609 Squadron carried out coastal patrols from Dungeness to Beachy Head, while 453 and 122 Squadrons carried out standing patrols between North Foreland and Dungeness. Both units were using Manston as a forward base.

10

The Fortunes of War

The weather on the first day of 1943 was overcast, with heavy rain showers limiting any flying or other activity on the airfield, although 609 Squadron carried out a small number of standing patrols with its Typhoons. There were no operational sorties flown that night and 609 Squadron's request to carry out a Rhubarb sortie was refused.

The following day was similarly uneventful; although 609 Squadron was scrambled several times throughout the day, nothing came of any mission and the unit resumed its standing patrols. A number of Albacores of 825 (FAA) Squadron arrived at Manston during the day to support those of 841 (FAA) Squadron that had arrived at the airfield during the previous August.

The first American Spitfire arrived at Manston on 13 January, when an aircraft from the 335th (USA) Fighter Squadron landed after its pilot had experienced engine trouble. The Spitfire's pilot, Lt Bishop, had been escorting B-17s during an operation at Lille, part of Circus 250. Unfortunately, he was not familiar with the airfield and one of the narrow legs of the undercarriage gave way as he touched down. A short while later, another American Spitfire from 336th (USA) Fighter Squadron landed successfully after his engine had failed over the Channel. The pilot, who was also returning from Circus 250, skilfully glided the Spitfire towards Manston and made a bumpy, but safe landing.

The 335th and the 336th Fighter Squadrons had previously been known as 121 and 133 RAF (Eagle) Squadrons, units that had been manned mainly by American volunteers. In early 1942, an officer, who would later be closely associated with RAF Manston, had taken command of 121 Squadron; this was Wg Cdr Hugh Kennard, who went on to form Invicta Airways. Both Eagle Squadrons had been returned to American command in August 1942, although the units still had many ties to the RAF. The third Eagle unit, 71 Squadron, was designated the 334th Fighter Squadron and, along with the other two, it then became part of 4th Fighter Group.

Flt Lt Jean de Selys Longchamps and his wingman, FS Andre Blanco of 609 Squadron, flew to Belgium on a daring personal mission on 20 January. The sortie began with an attack on a train near Bruges, where they were split up, with Blanco going after another train. Flt Lt de Selys Longchamps flew on to Brussels, where the former cavalry officer flew over and photographed the Palais de Justice, Palais-Royal, and cavalry barracks. Flt Lt de Selys Longchamps then climbed up to 2,000 feet and dived on the Gestapo headquarters on the Avenue Louise, raking the building with his four 20-mm cannon.

Flt Lt de Selys Longchamps then climbed back up to 2,000 feet and threw a Union Jack and the Belgian flag out of the cockpit, watching them descend into the centre of Brussels before flying over the area again to photograph the same building with the flags draped on the ground. It was a daring stunt that he had been planning for a number of months and on this day he judged the conditions to be just right, but he was still not finished. As he flew towards the coast, de Selys Longchamps flew over a number of villages, dropping hundreds of miniature Belgian flags that he had also carried in the cockpit of his Typhoon.

The two Belgian pilots landed back at Manston at 9.44 a.m., but de Selys must have been very disappointed to find out that his camera had not been working and none of the photos of his escapade had come out. However, it was later claimed that de Selys had killed at least thirty Germans during his raid on the Gestapo HQ, but also, sadly, a member of the resistance, who had been posing as a collaborator.

The 20th was generally a very good day for 609 Squadron, and, at 8.34 a.m., Red Section had taken off on a defensive coastal patrol. Flying over Dymchurch, they encountered two Fw 190s just inland of a convoy that was sailing past Dungeness. Fg Off. Lallemond dived on what he described as the No. 1 aircraft, firing a burst that caused flames to appear on the wing and in front of the cockpit. The other Fw 190 flew off in the direction France, but unnamed sources later confirmed that the aircraft had crashed into the sea. Red Section landed safely at 9.30 a.m.

At 12.29 p.m., another section of 609 Squadron was scrambled to intercept a number of Bf 109s that were flying around to the east of Manston. Fg Offs Baldwin and Creteurs found that there were no less than eight enemy aircraft and successfully engaged them, with Baldwin claiming to have destroyed two and damaged one. After the interrogation of a prisoner of war (a captured enemy pilot), it was later found that all three enemy aircraft had been destroyed. Fg Off. Baldwin had unwittingly set a new squadron record.

On 26 January, and after a short ceremony held by Wg Cdr Sheen, the station said 'goodbye' to the night-fighter flight, which had arrived in November 1941. It had been a mixture of experienced pilots from 3 and 151 Squadrons—among them had been Sgt Scott and the 'lone wolf', Flt Lt Stevens.

Under the command of Flt Lt Collins, the seven Hurricane IICs of the flight took off to fly to Hunsdon, where it was to be re-equipped. After getting airborne, the Hurricanes flew out to sea before returning to do a wide sweep of the airfield, dipping their wings as a salute to those airmen standing at their dispersal. In the fifteen months that it had existed, the Manston night flight was credited with destroying seventeen enemy aircraft, another four as probables, and sixteen damaged.

Wg Cdr Scott-Malden, DSO, DFC, from Fighter Command HQ, visited Manston on 12 February with a group of Royal Engineers, who were dealing with the rapid preparation of advanced airfields and landing grounds in the south of England. On the airfield, they investigated a number of sites described in the ORB as the 'notorious bumps and hollows' that were to be levelled out as they were considered a danger to modern types of aircraft.

This was a period of change, and Grp Capt. Mulholland of Fighter Command HQ and Grp Capt. Pike from 11 Group were also at Manston to inspect and discuss changes to a number of buildings. Those that came under scrutiny were the centralised armaments, servicing building, and the east camp cookhouse. Plans were also made to accommodate better messing facilities off the station in Manston village.

The ORB went on to mention that formulation of plans for an extension of the east–west runway was being pursued by the Air Ministry, but no detailed information as yet had been received. The much needed reinstatement of the perimeter track as requested by Wg Cdr Sheen was to commence in the near future, with a section of it leading up to the 'R and R' Hangar.

The fact that a number of skilled craftsmen had been called up was blamed for the reduction in the amount of maintenance work taking place on the camp, and it was noted in the ORB that there were only two plumbers available. It was claimed that it was only because there had not been any recent hard frost that many problems with the plumbing had been avoided. There was not the slightest doubt that a period of cold weather would have led to a series of burst pipes, which would have meant a lot of discomfort for all.

Fog was responsible for a serious accident on 19 February, when, at 1.10 a.m., two Westland Whirlwinds of 137 Squadron collided on the runway and both pilots were killed. One of the aircraft was taking off, while the other was taxiing and the bombs that they were carrying exploded immediately, with flames engulfing both aircraft. Both pilots came from countries of the Commonwealth and one of the pilots was twenty-one-year-old Lt Neville Austin Freeman (19862), son of Cyril E. and Charlotte B. Freeman from Cape Town, South Africa. A member of the South African Air Force, he had been flying Whirlwind P7119 (SW-F), which collided with Whirlwind P7114, flown by Plt Off. Charles Eldred Mercer of the Royal Canadian Air Force.

Right: Pilot Officer Charles Eldred Mercer of 137 Squadron, who was killed on 19 February 1942 when his Westland Whirlwind collided with another Whirlwind on the airfield at Manston.

Below: The line-up of 137 Squadron's aircrew with a Westland Whirlwind taken prior to June 1943, when it converted to the Hurricane Mk IV.

Flying Officer Eddie Ashfield of 137 Squadron posing by his Hawker Typhoon armed with rocket projectiles.

The runway was cleared by the morning and, at 7.15 a.m., a section of Typhoons from 609 Squadron took off, but were forced to return when fog threatened to engulf the airfield. Later the same day, an evasion exercise was carried out by 823 and 841 (FAA) Squadrons. Twelve pilots and observers from the units were taken to various points, approximately 6 miles from the airfield, and were left to find their own way back. To complete the task, they had to enter the airfield unobserved, with all but one of them completing the exercise; the one that failed was picked up by an armoured car belonging to the RAF Regiment.

The ARP, the police, and Home Guard were actively involved in the exercise and scoured the countryside searching for 'suspicious-looking' characters. The following day, a similar exercise was carried out with ten pilots from 609 Squadron, who were dropped at locations between Chislett and Reculver. Nine of the ten pilots achieved their aim and found the intelligence officer, while the tenth was apprehended while trying to enter the camp on a public bus.

The first day of March 1943 was fine, with a slight cloud cover, but vertical visibility was restricted because of haze. There was considerable enemy activity over the French coast and at one point the whole of 609 Squadron was brought to readiness, with one flight being scrambled, but ordered to land after only being airborne for a few minutes. That night, two Albacores from 823 Squadron, flown by Lt Douglas and Sub-Lt Stanley, carried out a shipping reconnaissance patrol between Dieppe and Berck. Visibility was between 3 and 10 miles, and there was no sign of any enemy activity.

There was another remarkable feat of airmanship at Manston on 4 March, made by Capt. Weisteen of 331 (Norwegian) Squadron. Flying over Calais, at a hight of only 2,500 feet, the engine of his Spitfire cut out, forcing him to glide the aircraft back over the Channel. He managed to reach Manston, but found that, even with his wheels down, he had too much speed for a conventional landing, so he raised the landing gear and force-landed the aircraft.

On the same day, a large party of the press visited Manston to pay an official visit to 137 and 609 Squadrons; the ORB noted that 'many stories of noble deeds were told'. It was suggested that, over the next few days, these would appear in various stages of distortion in the national newspapers. To keep the press happy, the Whirlwinds and Typhoons performed what was described in the ORB as 'hair-raising aerobatics'.

The following morning the airfield was besieged again by photographers from the press, who tried to take photos of 137 Squadron's Whirlwinds from every conceivable angle. Just after the photographers had gone, an incident occurred that would have made front-page news—had the censors allowed it—when a Typhoon of 609 Squadron was fired upon by the local AA battery. The aircraft was returning from an aborted sortie due to the hazy conditions, when the battery opened fire and engaged it for two minutes before being ordered to stop firing by Flying Control. Sqn Ldr Beaumont and the Station Intelligence Officer also intervened and someone at the AA battery got sent a 'rocket'.

On 9 March, 609 Squadron lost a Typhoon during air-firing practice, with the aircraft crashing a few miles south of the airfield near Great Knell Farm at Ash at 2.50 p.m. The Typhoon, DN481 PR-N, was being flown by a pilot that was new to the unit, Sgt N. Booth, and it was his inexperience on the type that was blamed for the accident.

That could not be said about Sqn Ldr Beaumont, who, just a couple of hours later, had to make a forced-landing near Deal after the engine of his Typhoon failed during a defensive patrol. The aircraft was badly damaged and the squadron leader suffered from concussion and was confined to his bed for several days. The third and final accident of the afternoon occurred just before dusk, when a Spitfire from the PRU overshot on landing and collided with a petrol bowser. The pilot was very lucky to have escape without being injured.

No. 609 Squadron lost another Typhoon on 25 March during a defensive patrol, when Fg Off. Baldwin's section had been attacked from behind by two Fw 190s that shot his controls away. He had heard someone shout 'Tally-Ho' over the R/T and presumed that the call was to warn him of approaching enemy aircraft, but he failed to act. His aircraft, DN560, was seen emitting black smoke and entering a spin; however, with considerable difficulty, Baldwin managed to bale out at 1,000 feet and he was lucky to get away with burns to his legs, eyes, and hands. Baldwin was soon picked up by a high-speed launch that was operating out of Ramsgate, but the skipper had put to sea without authorisation from Dover and so he was subsequently 'carpeted' for putting to sea without orders.

No. 198 Squadron arrived from Acklington with its Typhoons on 28 March, under the command of Sqn Ldr J. Manak. The unit had been reformed at Digby in December 1942 and its main role at Manston was to provide escorts for the Whirlwinds of 137 Squadron, which were operating in a fighter–bomber role.

No. 609 Squadron lost another Typhoon on the 29th, when, at 5 p.m., R8888 was seen entering the circuit with white smoke streaming from its engine, fast disappearing towards the east. The pilot, twenty-year-old Sgt Jackson, was trying to get into a position to land, but the Typhoon was seen to be on fire and it went out of control and crashed into an empty house, killing him. If it not already done so, the Typhoon was getting a reputation for engine and structural failures. There was no further operational activity that day at Manston.

At the end of March 1943, the total strength of RAF Manston was 1,979 personnel, with 100 RAF officers, and five WAAF officers, commanding 1,218 airmen and 196 WAAF other ranks. The resident units were 137, 609, 198, and 841 (FAA) Squadrons, with 4019 Anti-Aircraft Flight and 2778 and 2844 Squadrons of the RAF Regiment for the defence of the airfield.

11
A Distinguished Visitor

The first day of April 1943 was full of various types of activity. At 6 p.m., a Mosquito from 139 Squadron appeared in the circuit and it was obvious that the aircraft had been badly damaged by enemy action. With one engine feathered and having lost the use of his artificial horizon, gyro, turn and bank indicator, and TR9 radio, the pilot, Fg Off. Talbot, struggled against the odds to land safely.

Talbot later received a message from the AOC of No. 2 Group, who stated: 'I consider that you and your observer, Sgt Sleeman, put up an excellent performance in getting your aircraft back this afternoon. A very good effort by both of you'.

On 11 April, RAF Manston had a number of important visitors, including Wg Cdr Guy Gibson, who had recently been given the task of forming 617 Squadron. Gibson had arrived at Manston in a small single-engined, two-seater trainer—a Miles Magister—and the purpose of his visit was to observe the demonstration of Barnes Wallis' 'Bouncing Bomb' off the coast at Reculver. Later that afternoon, he departed to go back to the squadron's base at Scampton, accompanied by a colleague, Australian-born bomb aimer Flt Lt Bob Hay.

Just a minute or so after taking off from Manston, the 130-hp de Havilland Gipsy Major engine of the Magister failed, leaving the wing commander with very little height or time to choose where to put the aircraft down. Most open land and fields in the area were full of obstacles, which acted as anti-invasion measures, but as the Magister drifted across Birchington he was able to choose a spot. The wing commander skilfully made a forced-landing at Brooks End Hill, to the south-west of Birchington, between the villages of St Nicholas at Wade and Acol.

As it approached the ground, the aircraft was spotted by a thirteen-year-old boy, Ted Sturgess, who, together with his friend Harry Castle, ran towards where it had come down. The aircraft was wrecked and the boys were surprised when the two officers climbed out, shaken up and worse for wear, but not seriously injured. Neither of the boys recognised the airmen (there was no reason at this point why they should have), although they had seen the aircraft doing 'dummy' runs along the coast and practising for the raid that was to make history just over a month later. It was a number of years later that the two men found out who they had rescued that day.

A recovery team from Manston soon arrived on the scene to collect Gibson and Hay, although the aircraft was left in place for a number of days before being removed. In his logbook, Wg Cdr Gibson simply noted: 'Local Manston. Crashed in field. OK'. He also later mentioned the incident in his book, *Enemy Coast Ahead*:

> When we were at 300 feet over Margate the engine stopped. Anywhere else it would be easy to put the machine down in a convenient field but not at Hells Corner where they make certain that aircraft do not land safely in fields. There were abundant wires and other devices because German glider borne troops were not very welcome.

Although he was very busy making preparations for the Dams Raid, it was another four days before he flew again, this time with two crew in an Airspeed Oxford to Reading and Fairoaks. The incident involving Wg Cdr Gibson is not mentioned in the Manston ORB, but that may not be unusual as there was a lot of activity that day, as well as the fact that what had happened at Manston was classed as top secret.

At Manston, Barnes Wallis and his team had been allocated their own hangar that was closely guarded at all times, although it was claimed strange shapes described as 'large golf balls' had been seen through 'cracks' in the hangar doors. Wg Cdr Gibson stayed over at Manston on at least one occasion and was seen at breakfast in the old Officers' Mess at Pouce's Farm. There were so many scientists and civil servants at Manston that the CO, Wg Cdr Raphael, was allowed the rare award of an entertainment allowance, but warned to spend it carefully. At 6*d* a day, Wg Cdr Sheen thought that there was not much chance of that happening.

On 16 April, 609 Squadron lost another Typhoon, R7855, when it crashed after carrying out a defensive patrol between 1.36 p.m. and 2.40 p.m. The aircraft was being flown by Sgt Aitken-Quack, who was not badly injured, although the Typhoon was assessed as being damaged Cat 'B'. The reason for this incident is not known, but it does state in the ORB that the aircraft had not been damaged by enemy action.

Wg Cdr Gordon Learmoth Raphael took over as the Commanding Officer on 18 April from Wg Cdr Sheen, who was posted to the Air Ministry. Raphael had risen through the ranks quite rapidly. He had originally been a bomber pilot and had flown Whitleys with 77 Squadron, but, having survived a ditching in the North Sea and spending four hours in the water, he had transferred to Fighter Command.

Raphael had then joined 43 Squadron and flew Hurricanes from Drem, but, by the end of 1941, he had been posted to 96 Squadron at Cranage in Cheshire. With the rank of flight lieutenant, he was the Flt Cdr of 'B' Flight, flying the Hurricane and Defiant in the night-fighting role. He had then been promoted to squadron leader and posted to 85 Squadron at Hunsdon, where he was later promoted to wing commander and had taken over as its Commanding Officer. Raphael, who was a Canadian and a 'teetotaller', had not been one of the most popular officers in either 96 or 85 Squadrons, especially after he had tried to control the drinking habits of other officers and airmen—although without much success.

Manston continued to be regularly used by aircraft from Bomber Command and, on the night of 26 April at 11.19 p.m., a Mosquito from 105 Squadron crash landed on the airfield—the aircraft had been hit by flak over Tours in France. Its port wing tip, engine nacelle, and fuselage were badly damaged, and the pilot had lost all his instruments, electrics, and hydraulics. At 3.52 a.m., a Whitley from 24 OTU landed after the pilot had got lost after losing the artificial horizon and a number of other vital instruments. A third aircraft, a Wellington from 466 Squadron at Leconfield, landed at 4.52 a.m., short of fuel after bombing Duisburg. The crew were convinced that their aircraft had been hit and damaged by flak, but an inspection of the airframe proved the aircraft had not been hit.

A few days after making his forced-landing at Birchington, Wg Cdr Gibson was back in the training programme. On the 16th, he was flying Lancaster ED593 on a low-level exercise involving dummy attacks on various lakes. On 27 April, Gibson returned to Manston in Lancaster 'D', but again this was not mentioned in the ORB. The reason for his visit was almost certainly to observe another test dropping of the 'Bouncing Bomb' at Reculver as there were still problems with the casing of the weapon.

On 1 May, Gibson flew to Manston in Lancaster 'H' with his crew and Sqn Ldr 'Dinghy' Young. It was at this point that there was a security scare when Gibson discovered that an armaments officer, Plt Off. 'Doc' Watson, was aware of the operation and had been shown the plans. Watson had spent three weeks at Manston during April, in work connected to the bombing trials. Back at Scampton, he had revealed to Gibson that, just a few days after arriving at Manston, he had been shown a file that contained detailed information, which included diagrams and maps about Operation Chastise.

As one can imagine, Gibson was furious about the 'Manston File' and the fact that it contained such detailed information that even he had not seen; when he found out that an officer from 618 Squadron (618 Squadron had

Annotations on top photo:
- SGT. N. JACKSON V.C
- BOB KNIGHT — Bomb Aimer
- BILL HOVEY — Navigator
- VAN STOCKHAM — Mid Upper Gunner
- WHI... Rear g...
- LES LAWLAN — Flight Engineer
- JOCK CUNNINGHAM — Pilot

Wing Commander Guy Gibson saying farewell to airmen and officers of 106 Squadron at Syerston in March 1943 after completing seventy-one operations. After being given command of 617 Squadron, he visited Manston on a number of occasions in preparation for the Dams Raid. Note the Avro Manchesters in the background.

Above: Avro Lancaster ED593 ZN-Y *Admiral Prune II* at Metheringham in April 1944, used as a backdrop for this group photo of 106 Squadron aircrew. Gibson had been the CO of 106 before he went on to form 617 Squadron; he was a regular visitor at Metheringham, often meeting old friends. The aircraft was used by 617 Squadron during early trials and dummy runs and landed at Manston on at least one occasion on 7 May. The photo features Sergeant Norman Jackson, who was awarded the Victoria Cross for his actions on 26 April 1944—just a few days after it was taken.

been formed on 1 April with Mosquitos to use a similar weapon called 'Highball' for an attack on the *Tirpitz*) had seen the file, it did nothing to improve his mood. When Gibson complained to the AOC of 5 Group, Air Vice-Marshal Ralph Cockrane, he agreed with him that the armaments officer should not have been allowed access to the file.

The officer at Manston who had shown Watson the file had it immediately removed from his possession and was severely reprimanded for threatening the security of the operation. Gibson was at Manston again on 7 May, flying Lancaster 'Y' with what he noted as his usual crew. The following day he was flying in a Vega aircraft with Flt Lt Cox, arriving from Hendon and departing for Scampton. That was Wg Cdr Gibson's final visit to RAF Manston.

At 7 p.m. on 13 May, four Spitfires from 453 (Australian) Squadron arrived at Manston for a special operation—to escort a Lancaster that was to lay a 'mine' in the estuary. Two of the Spitfires escorted the aircraft and the other two escorted another Lancaster that was carrying the Secretary of State for Air and a number of Senior Staff officers, who were there to observe the operation. It was noted in the ORB that the operation was successful and afterwards the four Spitfires flew back to Hornchurch where they landed at 8.20 p.m. This was in fact the final trial of the bouncing bomb, and it was carried out 5 miles off the coast of Broadstairs, with a live 'upkeep' weapon dropped exactly as Barnes Wallis had ordered—from a height of 60 feet and at a speed of 232 mph.

After leading the unit for six months, on 5 May, Sqn Ldr Beamont handed over command of 609 Squadron to Sqn Ldr A. Ingle. On the 14th, news was received at Manston that the King, on the recommendation of the AOC of Fighter Command, was awarding the DSO to Sqn Ldr Beamont, DFC and Bar. A tribute to him in the Manston ORB stated:

The immense keenness and the very great spirit of the squadron, was very greatly improved as a result of his immense activity on behalf of the squadron. He had, moreover, shown that properly handled, and properly run, the Typhoon was about the best type of aircraft for the combating of the present enemy policy of low-flying raids of short penetration.

Promoted to the rank of wing commander, Beamont was returning to the Hawker company where he was to work on an advanced version of the Typhoon, the Hawker Tempest.

Also on the 14th, a signal was received at Manston that stated 198 Squadron was to move to Woodvale, on the west coast of Lancashire, with immediate effect. The problem was that most of the pilots and aircraft were not at Manston, but at Martlesham on an air-firing exercise. Nevertheless, those airmen and officers on the detachment were ordered not to return to Manston, but to proceed directly to Woodvale. It was arranged that those elements of the unit that had remained at Manston would be responsible for packing up the personal kit and effects of the other members of the unit and transporting it to Woodvale. If things were not difficult enough for the squadron, in the middle of the move, the squadron adjutant was posted to another unit; however, when it was realised what was happening, he was given an extra three days at Manston to complete things before moving on.

On the day of the Dams Raid (Sunday 16 May), Manston was very busy. It began as a fine day, with 2/10ths cloud and various amounts of cloud cover later on. Four Typhoons of 609 Squadron took off at 5.15 a.m. on a Roadstead operation to Flushing and Ostend. Among other movements was the arrival of two Republic P-47 Thunderbolts from USAAF's 84 Squadron, which was based at Duxford. One of the pilots of the Thunderbolts had been wounded in combat. The pilots were 2Lts Barba and Dowling, and it was the former who had been badly wounded in his left arm and was suffering from shock. Barba's Thunderbolt was not a write-off, but it was assessed as Cat 'B' damage and he was taken to Margate Hospital where a number of bullets were removed.

That night, 137 and 609 Squadrons were involved carrying out 'searchlight box patrols' over Canterbury, but there were a lot of problems of identification by the AA battery that could easily have led to a disaster. At 12.05 a.m., a Whirlwind of 137 Squadron took off and it was over Canterbury at 7,500 feet by 12.40 a.m., when it was effectively 'coned' in the searchlights for fifteen minutes. During that time, it was fired at by both the light and heavy guns of the AA defences, but fortunately their shooting was not accurate and they failed to find their target. The Whirlwind landed safely at 1.15 a.m. with a very unhappy pilot, who was about to file a report about the incident.

Fg Off. Lallemende took off to carry out the same patrol at 12.23 a.m., and over Canterbury in the searchlight box area he spotted three enemy aircraft. He was unable to do anything about it because at about the same time he was caught and held in the searchlight beam. Despite repeatedly flashing the colour of the day, the ground defences failed to respond. Lallemende eventually managed to get away from the searchlights and landed safely at 1.44 a.m. It was stated quite clearly in the Manston ORB that both these aircraft had their IFF (Identification Friend or Foe radar aid) switched on all the time that they were being fired at.

The Dams Raid is mentioned in Manston's ORB, but the entry comes under the heading of 'General Information' and is not dated, although it was obviously written after the event. It states that a lot of work had been carried out in the Bellman hangar under the greatest secrecy, and that Wg Cdr Garner had been in charge of the detachment from the Ministry of Aircraft Production. It goes on to say that the Secretary of State for Air, Sir Archibald Sinclair, had lunched at Manston on Sunday 16 May and the following day, when news was released that Wg Cdr Gibson had led the raid on the Dams. Whoever wrote the ORB observed that what had gone on at Manston in great secrecy must have been connected with the operation, otherwise it was must have been a great coincidence. Gibson himself noted in his logbook the sparse comments, 'Led attack on Mohne an Eda Dams. Successful [*sic*.]'.

Although both Gibson and Hay, who had survived the incident with the Magister on 11 April, returned safely from the Dams Raid, neither of them survived the war. Flt Lt Robert Claude Hay was killed in action in February 1944 while still flying with his old crew, led by Flt Lt Micky Martin. Having gone on to become the bombing leader of 617, Sqn Ldr Hay was killed by a single shot fired from the ground while flying over Italy.

As is common knowledge, Wg Cdr Gibson was killed on 19 September 1944 while returning from a bombing raid, on which he had been acting as the master bomber (Master of Ceremonies). One can only muse over what would have happened to the Dams operation if Gibson had been killed at Birchington in April 1943 and not at Steenbergen Holland in September 1944.

12

Focke-Wulf Galore

What was described in the ORB as a 'Grand Finale' to the moon period occurred on 20 May, when, at 3.40 a.m., a Fw 190 from 2/SKG 10, call sign *Red 9*, landed at Manston by mistake after its pilot, *Unteroffizier* Heinz Ehrhardt, had got himself lost and had confused the Thames Estuary with the English Channel. Both the local searchlight battery and Flying Control Officer, Plt Off. Dreyer, were congratulated for their part in capturing the enemy aircraft. The searchlights for homing the aircraft and Dreyer for his considerable ingenuity in working the airfield lighting, which had actually been switched on to aid a 609 Squadron Typhon, flown by Flight Lieutenant Roy Pain, in landing, were commended for their efforts.

A number of 609 Squadron pilots, including Squadron Leader Ingle, apprehended Ehrhardt before he could react and he was taken to the Intelligence Office. That was where Flight Lieutenant Paine later found him, being given mugs of coco and offered cigarettes. Eventually, Ehrhardt was put in the charge of Flight Lieutenant Keen from the Intelligence branch AI (K), who informally interrogated Ehrhardt before he was taken up to London to undergo more rigorous questioning.

The Fw 190 was handed over to Flight Lieutenant Clowes from another Intelligence branch, AI 2 (G) and it was soon discovered that the aircraft's tail wheel had been damaged on landing. Pilots of both 137 and 609 Squadrons were somewhat disappointed that they would not get the opportunity to fly the aircraft. There was more bad news when it was announced that the Fw 190 could not be repaired at Manston and it would have to be taken away and dismantled. The Fw 190 later became part of 1426 Flight that evaluated captured enemy aircraft; having been allocated RAF serial number (PM679), it continued to fly until June 1944, when it was written-off after a crash.

There was a follow up to this story, when, after the war, Squadron Leader Roy Paine was in Germany on a training programme working with former Luftwaffe pilots. The squadron leader was telling his German opposite number the story about the Fw 190 and what happened on that day. The German pilot suddenly disappeared before returning a few minutes later with his logbook and, checking the dates, he found that he had been flying with Ehrhardt the night that he had gone missing. As far as he was concerned, Ehrhardt had gone missing and his fate had been something of a mystery as to whether he had been killed or taken prisoner. Only after his conversation with Roy Paine many years later did he discover what had happened that night to his leader.

By the end of May 1943 it, had become evident that work on the building and extension of the new east–west runway, which was to be 3,000 yards long, was progressing well. The resident engineer and the clerk of works had met to discuss the plans and, at the end of May, a representative from the contractors was based on the station. The contract had been given to Messr Laing, a company that had been involved with many similar projects and had a huge amount of experience in building runways on RAF stations.

June began with a hectic start, and, during the morning of the 1st, 609 Squadron was active on Roadstead sorties off the Dutch coast near Walcheron and Schouwen. There, Squadron Leader Ingle and Norwegian Flight Lieuyenant Haabjoern led a determined attack on local shipping and claimed to have seriously damaged or sunk a total of five vessels. These included a 500-ton flak ship, a flak tender of between 200 and 400 tons, and two other vessels, whose estimated tonnage was 800 tons.

At 12.20 p.m., two Typhoons of 609 Squadron were in action again when they were scrambled after reports that enemy shipping was active off the French coast. The two pilots, Squadron Leader Wells and Flying Officer Davies, failed to find anything, but, while approaching Manston, flying at 1,800 feet and still under the control of Hornchurch, they spotted three enemy aircraft flying in Vic formation. The aircraft were Fw 190s, flying from the direction of Margate towards Broadstairs; with thick black smoke rising into the air, the airmen realised that they had caused a lot of damage. They later found out that Margate's gasworks had been attacked and had literally gone up in smoke.

A few moments later, another twelve enemy aircraft were spotted and it was soon realised that this was a major operation by the Luftwaffe, with all the aircraft flying the same course of 110 degrees at zero feet. It was later described as a 'terror raid' because the Fw 190s were not just dropping bombs, but machine gunning the streets to cause as much death and destruction as they could.

Squadron Leader Wells reacted quickly and he soon caught up the with the Fw 190s; within a short time, he shot two of them down into the sea before suffering great disappointment when he had the third one in his sights. When he opened fire, he discovered that he had run out of ammunition, but, to add insult to injury, his aircraft was hit and damaged by the AA guns as he flew over Broadstairs, forcing him to return to Manston.

As Wells turned back, he saw a number of other Fw 190s passing beneath him and he saw that they were being chased by Flying Officer Davies, who almost immediately shot one of them down. Davies then went on to destroy another two Fw 190s, while the AA guns accounted for one as well, making a total of six enemy aircraft shot down. Apart from those claimed by 609 Squadron's own pilots, the Dover AA battery confirmed that four enemy aircraft had been destroyed, having seen them crash into the sea.

The pilot of the Fw 190 that was shot down by the local AA defences was *Unteroffizier* Otto Zugenbrucker, who had baled out, but was killed when his parachute failed to deploy. Flying Officer Davies' record of destroying three enemy aircraft equalled that of Flying Officer Baldwin after he had done the same thing on 20 January. However, as the Manston ORB pointed out, Baldwin's record still stood because he had done it in a much quicker time.

On 4 June, Manston had a visitor from North Africa. At 12.27 p.m., a Halifax landed without any prior notification after what was described as an uneventful journey from North Africa. The only thing of interest to be reported was that, at 3 p.m., off the coast of Lisbon, it had spotted a large vessel of approximately 10,000 tons heading west-south-west. The aircraft had taken off at 12.50 p.m. from Sale in French Morocco and was based at South Holmsley in Hampshire, to where it returned the next day.

On 7 June, what was described as a 'report on operations at RAF Manston' mentioned an incident involving a Halifax that had landed on 30 May with a 'hang up'—a 1,000-lb bomb that had been dropped live, but had failed to release itself from the aircraft's bomb bay. The report noted that the bomb was a 1,000-lb general purpose weapon that had been fitted with a Type 845 Mk II nose and a 37 Mk IV tail (six-hour delay). The bomb had been removed from the Halifax under the direction of the station armament officer, Pilot Officer Spencer, and was quickly secured, segregated, and taken to the bomb dump. A further inspection of the bomb confirmed that the fuse on the bomb had not been armed.

The report went on to say that, because the bomb had not been jettisoned or recovered from a crashed aircraft, the incident did not come under the scope of the Bomb Disposal Organisation. The bomb could not be returned to service because of the type of fusing fitted and because it had been separated from the fuse setting control links. Requests had been made to both the aircraft's parent unit and Fighter Command to ask for suggestions to deal with this; the only two ideas put forward were to either load the bomb onto another aeroplane and drop it out to sea or to ask the Royal Navy to deal with and dump it into deep water.

If neither of those options were acceptable, then the Air Ministry would accept responsibility, but it was understood that the bomb would have to be demolished by detonation. For that to be arranged, there had to be an imposition of a 1,000-yard safety area, but at Manston that might present a hazard and interfere with flying operations since a bomb so fused would not be allowed to be transported along public highways without an evacuation taking place.

During the morning of the 7th, the station armament officer at Manston rang the Air Ministry to confirm that he had been unsuccessful in clearing the bomb, but requested assistance from the Air Ministry. Squadron Leader Scamells and Flight Lieutenant Wilson left the Air Ministry at 2 p.m., arriving at Manston at 4.30 p.m. and made contact with No. 8222 Bomb Disposal Flight. Aided by Pilot Officer Spencer, a search was carried out where the bomb could be safely dealt with at Manston and, after two or three hours, a suitable location was found on the far side of the airfield. There were still problems, however; if the site was chosen, the barbed wire fence that formed part of the station's defence had to be cut down and all flying operations stopped.

It was at that point that the Station Commander, Wing Commander Raphael, became involved and suggested that a former 'underground' hangar, which had been built during the First World War, should be used. It was close to the bomb dump and another advantage was that it was some 20 or 30 feet below ground level and so able to deflect the blast more easily. With the help of the local home guard and the police, only a single building needed to be evacuated and a road closed before the bomb was moved to the site at 8 p.m. By 9.10 p.m., using tools and equipment that had been developed on site, the bomb had been defused and was taken away by the bomb disposal unit.

The Anglo-American Bomber Offensive was issued on 10 June 1943 and was code named Pointblank. This offensive had been agreed between Roosevelt and Churchill at the Casablanca conference. Its aim was the destruction of German air power in preparation for an invasion of Europe in 1944. Although the Americans had been bombing enemy occupied Europe since July 1942, Pointblank would ensure that the USAAF's bombing raids would become more intensive and that Manston would play its part in its operations.

During the morning of the 11th, 3Squadron arrived from West Malling, equipped with the Typhoon 1B and under the command of Belgian Squadron Leader de Soomer, who had taken over in August 1942. There also were a number of other Belgians among 3 Squadron's pilots, but none so well-known as Flying Officer de Selys Longchamps, who had recently been posted in from 609 Squadron. To celebrate their return to Manston, four Hurricanes from 3 Squadron performed a 'beat-up' of Herne Bay as part of the station's contribution to help raise money for 'Wings Week'.

No. 137 Squadron left Manston for Southend on 12 June to exchange its Westland Whirlwinds for the Hurricane Mk IV, but some of its pilots found it difficult to keep away. The unit's replacement was 184 Squadron, which had only been formed at Colerne in Wiltshire in December 1942, arriving from Merston, in Sussex, equipped with the Hurricane. On arrival at Manston, its Commanding Officer, Squadron Leader Rose, was heard to make a comment over the R/T that the airfield looked like a 'pig's breakfast'. The remarks were not taken lightly by Wing Commander Rapahel, but things were about to change.

During the early hours of 20 June, a bomb was dropped south-west of the airfield by a Fw 190 and, an hour later, at 3.30 a.m., to everyone's surprise, another Fw 190 landed at Manston, almost exactly one month after Ehrhardt's Fw 190 had arrived. The pilot of this second one was *Unteroffizier* Werner Ohne, who had been attracted to Manston by the airfield beacon, which had been flashing the same letter as the one that he had expected to see near his base in France, close to Cape Gris Nez. Squadron Leader Wray, who had recently taken over command of 137 Squadron, was the first to reach the Fw 190, sticking a pistol in Ohne's ear before dragging him out of the cockpit before he could react.

Squadron Leader Mike Bryan, a Flight Commander on 137 Squadron, was about to land after returning from a sortie, during which, purely by chance, he had attacked the airfield from where the Fw 190 was based and he had followed the enemy aircraft around the circuit at Manston. When the Fw 190 landed, it was effectively blocking the runway—Squadron Leader Wray decided to move it so that Bryan and other pilots could land safely. Wray was a bit nervous at first, suspecting that the enemy aircraft might have a device to destroy it if any unauthorised person tampered with it, but his fears were unfounded. After taking lots advice from other pilots and still fearing that he might inadvertently raise the undercarriage, he pushed the throttle forward and successfully moved the Fw 190 out of the way.

It did not take long to discover that the captured enemy aircraft was one of the latest types and, being in near perfect condition, it attracted a lot of attention from the pilots of 137 Squadron. The following day, a number of officers from Intelligence Unit AI 2G arrived at Manston to examine the aircraft and interrogate the pilot. After a few days, the aircraft was flown out of Manston and it later joined 1426 Flight, which was the RAF unit that evaluated captured enemy aircraft. It was given the RAF serial number PN999 and eventually ended up at 47 Maintenance Unit, based at RAF Sealand, on the Welsh border.

13

Emergency Runway

An Air Ministry conference was held at Manston on 22 June 1943 to discuss the plans for the proposed emergency landing ground (runway), with Wing Commander Powell from the Air Ministry as the chairman. There were twenty-three people at the conference, at which there appeared to be some anomalies, with a Mr Mayo-Wells being listed as the CO RAF Manston and Wing Commander Raphael as a representative from the Air Ministry. There is no explanation for this other than that their roles at the conference were only regarding the emergency landing ground and not their current positions.

The Full List of those Attending the Conference

Wing Commander Powell. Air Ministry 0.9 (In the Chair)
Air Commodore Darley, OBE, Air Ministry
Squadron Leader Harden, RAF Manston
Squadron Leader McIvor, 11 Group
Squadron Leader Mills, Bomber Command (Air Staff)
Flight Lieutenant Broadhead, ORS, Bomber Command
Mr Mayo-Wells, OC, RAF Manston
Wing Commander G. L Raphael, Air Ministry
Mr D. A. Rossiter
Mr Green, Air Ministry
Mr Jennings, Air Ministry
Mr Tomlinson, No. 13 Works Area
Mr Tandey, Manston
Squadron Leader Crawshaw, HQ Fighter Command
Wing Commander Willis, Air Ministry
Wing Commander Allerston, Bomber Command (signals)
Wing Commander Vincent-Miller, HQ Fighter Command

The original proposal for emergency runways near the east coast of England for the use of Intruder activity and bad weather conditions had first been discussed at a Group Commanders Conference, which had been held at Bomber Command headquarters on 3 October 1941. The Commander-in Chief, Air Marshal Sir Richard Peirse, had decided to take up the proposal, giving it the highest priority, and groups within Bomber Command had been asked to forward their views on their requirements as soon as possible. RAF Manston was to be the third airfield to be constructed, the other two being RAF Carnaby in Yorkshire and RAF Woodbridge in Norfolk. It was estimated that the emergency runway at RAF Manston would be in operation by 1 May 1944.

A decision was made that proposed that the emergency runway should be 3,000 yards long and 250 yards wide, a bold move considering that most runways on RAF stations at this time were no more than 2,000 yards long. It was also agreed that there should be a grass undershoot/overshoot area that was 500 yards long and 250 yards wide—this would properly adjoin the runway. As Bomber Command was going to be the principal user of the runway, it was noted that it would be responsible for its operational development.

The Primary Functions of Emergency Runway

(a) To provide emergency landing facilities for Bomber Command aircraft returning from operations in bad weather.
(b) For use by aircraft in distress.
(c) To be used for diversions by Central Flying Control Bomber Command under special circumstances only, including Intruder activity at home bases.
(d) For one way landings only, i.e. East to West.

Secondary purposes of Emergency Runway

(a) Despatch of aircraft on long distance reinforcement flight.
(b) Strategic concentration of aircraft for operational purposes.
(c) BA Training.
(d) Use by aircraft of all commands (including USAAC) in emergency the same as any other Bomber Command station in similar circumstances.
(e) The possible use of the runway for glider operations had been suggested but a decision had not been made until further investigations had been made.

The approach to the new runway was to be east–west and Wing Commander Allerston from Bomber Command pointed out that there was no difficulty in finding a suitable position of the inner marker beacon for the beam landing system. However, the site for the outer marker beacon might present something of a problem because of the proximity of the coast and the town of Ramsgate.

A number of changes had to be made on the airfield to accommodate the new runway, including the demolition of the water tower on the eastern side of the airfield and the resiting of a AA gun position. There was no shortage of water and a 7-inch water pipe, which had been ducted, was to be built under the runway and Margate Corporation were to be consulted in case they wanted it diverted. Additionally, there was a water main at the eastern end of the runway that would definitely have to be diverted. There were changes to a number of roads around the airfield and it was mentioned that the road running from the west end of the runway to a junction at the south end of the runway, where it joined the main Canterbury to Ramsgate road, had already been closed.

The new runway was fitted with a rather complicated system of lighting and it had three distinctive coloured channels, with red lights on the south side, white down the centre, and green on the north side. The lighting on the runway was divided into 1,000-yard sections, so that in the event of an accident each section could be isolated. There were four lines of airfield lighting at every 100-yard interval and four lines of contact lighting every 50 yards. A bar of white lights across the runway indicated to the pilot that there was 800 yards of prepared surface remaining. For those aircraft that needed to make a belly-landing, a Glim flare path was laid on the grass runway north of the main runway so it would not be blocked.

Provision was to be made for a total of twelve salvage bays, but it was pointed out that the edge of the runway and some of those of the salvage bays would be positioned close to the edge of the Canterbury Road. Where they ran the closest to the road, arrangements were to be made for controlled areas to be established.

The new runway was certainly needed as, two days after the conference, the airfield was 'invaded' by fifty Republic P-47 Thunderbolts from the USAAF's 56th Fighter Group. The group comprised of the 61st, 62nd, and 63rd Fighter Squadrons that had sailed to Britain on the *Queen Elizabeth* in December 1942. They had landed at Manston to refuel and for its pilots to be briefed for an important Ramrod operation, which was then cancelled. Despite the cancellation, the diversionary operation involving eight Typhoons of 3 Squadron and nine from 609 Squadron went ahead and they bombed their objective, which was a Naval Maintenance Unit at St Omer.

A heavily escorted enemy convoy was reported passing through the Straits on the 29th and 609 Squadron was scrambled. At 11 p.m., they found the convoy and claimed a single E-boat as damaged. When 3 Squadron was scrambled a short while later, it failed to find any trace of the convoy and so six Albacores of 841 (FAA) Squadron took off and joined the search. Although three were recalled, the others claimed to have damaged a merchant vessel and an E-boat. In connection with the enemy convoy, the Germans opened fire with their long-range guns, firing across Channel—some shells landed very close to Margate and Ramsgate.

On the last day of June 1943, information was received that there were two large ships being escorted by a flak ship off the coast of Dunkirk. Eight Hurricanes from 184 Squadron and eight 'Bombphoons' of 609 Squadron that had been on standby to attack enemy airfields in France were diverted to attack the ships. When the Hurricanes arrived over the target area, it was found that the ships were too close to the harbour, where the threat from heavy flak and AA guns was considered to be too risky, and so they decided to return.

Just a short while later, a number of Typhoons from 3 Squadron arrived over the target area and decided to show the Hurricane pilots how it should be done. They dived straight into the flak and claimed to have destroyed a ship of between 1,500 and 2,000 tons, an E-boat, and a tug. They did not get away unscathed, however, and Pilot Officer Purdon was shot down; despite an intensive search by ASR vessels between Deal and Calais, there was no sign of him or his aircraft.

On the first day of July, 3 Squadron attacked enemy shipping off the Hook of Holland, but the other two units that were to carry out the operation, 184 and 609 Squadrons, turned back because of their limited fuel endurance. Flight Lieutenant Colling, DFC, led 3 Squadron into the attack and a number of ships, including a 1,600-ton vessel and one of 1,300 tons, were sunk and several others were damaged. Unfortunately, 3 Squadron lost three pilots—Flying Officer Little, Pilot Officer Benjamin and Sergeant Lawrence were shot down into the sea. Although Little was seen climbing into his dingy, air-sea rescue failed to find him.

No. 3 Squadron was sent out again later that day to bomb Courtrai, but they were recalled after crossing the English coast. The Manston ORB gave credit to all the pilots who had to weave around various excavators and workmen when they landed, as work to level the landing ground continued.

On the 4th, the German's long-range guns opened up across the Channel again and fired twelve shells in the direction of Ramsgate, with two of them seemingly bursting in mid-air. The Manston ORB notes that there were a number of causalities, but it fails to mention whether these were civilian or military. There was also an air of frustration because, despite repeated attempts to organise offensive patrols, all the plans had fallen through and there were no operational sorties from the station this day.

Locks and lock gates were the targets for 3 Squadron on the 11th, with Flying Officer Foster and Flight Sergeant Crisford attacking and destroying the lock gates on the Yprellee Canal with bombs and cannon. The following day, Flying Officer de Selys Longchamps attacked the lock gates at Bossuyt and, although he found his target, his bombs bounced and missed his objective.

At 7.40 a.m. on 14 July, a Boeing B-17 Fortress from the USAAF 535th Bomb Squadron, serial number 42-3211, 381st Bomb Group, based at Ridgewell, Essex, made a belly-landing at Manston after being badly damaged and fired at head on by a Fw 190 over its target airfield at Amiens-Glisy. The wing of the Fw 190 had struck the inner starboard engine of the B-17 and sheared off against the waist gunners position, also damaging the starboard stabiliser, but totally destroying the enemy fighter.

Rather amazingly, neither the pilot, Lieutenant Manchester, or any of his crew were injured, despite the fact that pieces of the enemy aircraft had penetrated the empty bomb bay and other parts of their aircraft. During the afternoon, another B-17 flew in from Ridgewell to collect the crew and fly them back to Ridgewell, but unfortunately, the aircraft burst a tyre while landing at Manston and both crews had to return to Essex by road.

Lieutenant SMP Walsh, who had just been awarded the DSC, took over command of 841 (FAA) Squadron in July and a short while later a number of his aircraft and crews left Manston for Exeter, from where they were to operate a small number of Swordfish and Albacores. It was noticed that one of the Albacores taking off from Manston had a motorcycle slung underneath it instead of the customary bicycle. That became the butt of a number of jokes from their RAF colleagues, who retorted that mechanisation was finally sweeping through the Navy in the form of motorcycles being laced to a staggering 'string bag'.

14

Modern Manston

The AOC of 11 Group Fighter Command, Air Vice-Marshal Hugh William Lumsden Saunders, arrived at Manston on 20 July 1943 to decorate two Belgian pilots, Flying Officer Remy Van Lierde of 609 Squadron and Flying Officer de Selys Longchamps, who was by this time with 3 Squadron. de Selys Longchamps had been awarded the DFC, despite the fact that he had upset some of the Belgian authorities by carrying out his unofficial sortie to Brussels in January. Flying Officer Remy Van Lierde was probably not as well-known as de Selys Longchamps, but, on 14 May 1943, he had been the first pilot to drop bombs from a Typhoon.

After the ceremony and at the request of the Belgian authorities, 609 Squadron was released from operations so that its pilots could attend a party at the Savoy Hotel in London. The party was held not just to celebrate Van Lierde and de Selys Longchamps being awarded the DFC, but for Belgian National Day and it was suggested in the Manston ORB that some pilots might have had other invitations in mind regarding their female friends. It was an appropriate time for celebrations because a couple of days later 609 Squadron was, after nine months, going to leave Manston and be replaced by 56 Squadron, which arrived on 22 July equipped with the Typhoon 1B from Matlask. No. 609 Squadron moved out on the same day and replaced 56 Squadron at Matlask.

It was noted in the Manston ORB that during July considerable progress had been made with the initial construction of the east–west runway, with contractors' huts and petrol pumps having been erected on the southern boundary of the airfield adjoining the Canterbury Road. It was also noted that large drainage trenches, extending the full length of the runway, had appeared with great rapidity.

What the author of the ORB did not know was that the trenches were being dug for the FIDO system, which was still a secret. The work on that was being done by a different company to Laings, who were building the new runway, and the FIDO contract had been awarded to John Leoanard Eve & Co. Work had also commenced on the north-east to south-west runway on 10 July, with 300 men using grading and excavating machines under the command of Major McLeane. One of the depressions had been tackled by the borough engineer for Margate Corporation, while the others had been dealt with by the Army. Large mounds of earth had suddenly appeared on the runway, but a landing lane had been laid out on the eastern side of the airfield.

In the month of July, there were 383 operational sorties by the squadrons based at Manston, as well as 792 training sorties. The airfield was also used by 526 visiting aircraft for operational sorties. Additional Bofor guns were allocated the station and a number of those already there were resited, with earth taken from the levelling of the airfield being used to construct mounds to improve their field of fire.

According to the ORB, 1 August was the quietest day as regards operational activity for some time and although two Albacores from 841 (FAA) Squadron flew reconnaissance sorties during the early hours of the 2nd, they found nothing to report. The first visitor the following day was a B-26 Maurader of USAAF's 455th Bomb Squadron, flown by Lieutenant Kahley. The aircraft had been hit and damaged by flak while bombing the airfield at Merville and, although none of the crew had been injured, Kahley chose to land at Manston to check that his aircraft was still safe to fly.

There were a lot of comings and goings in August. On the 5th, 164 Squadron arrived from Warmwell, under the command of Squadron Leader McKeown, AFC; the unit was equipped with the Hurricane IV. The following day, 197 Squadron arrived from Tangmere with its Typhoons and it was commanded by Squadron Leader Holmes. No. 197 Squadron was another relatively new unit, having only been formed in November 1942. It had been attached to Manston to stand in for 56 Squadron, which had been sent on a fortnights' air-to-air firing exercise at Martleshem. During this period of the change, the state of readiness was maintained by 3 Squadron.

Having been re-equipped with the Hurricane IV to replace its Whirlwinds, 137 Squadron returned to Manston on 8 August, although some of its pilots had found it difficult to stay away and had regularly visited the station to have lunch. Flight Lieutenant Smith of 609 Squadron ditched his aircraft off the coast of Deal during the afternoon after flying what was claimed to be the first operational sortie in a Typhoon over Germany. The aircraft had been fitted with long-range tanks, but had not carried enough fuel to get him back to Manston. Smith suffered some minor burns, which suggests that his aircraft had been on fire, but regardless of his injuries he was fit enough to swim ashore.

At 3.30 a.m. on 11 August, the seven-man crew of Lancaster, serial number JA931, had a lucky escape when the aircraft ran out of fuel soon after crossing the Kent Coast. The Lancaster that was being flown by Flying Officer A. J. Belsey belonged to 7 Squadron, based at Oakington, and it was returning from an operational sortie to Nuremburg. According to the Manston ORB, the crew baled out over Birchington, where the Lancaster was reported to have crashed in an open field. The crew were then taken to Manston, where they were debriefed about the operation and how they had abandoned the aircraft. However, another version of events claims that the Lancaster crashed 10 miles north-east of Canterbury, the city being some 15 miles from the village of Birchington. In many ways, this explanation makes more sense because the Lancaster would have to have been at least 1,000 feet high and flying straight and level for all seven of the crew to bale out safely. That means it would have continued in flight for at least another minute or two, and it therefore seems quite likely that the crew baled out over Birchington and the Lancaster crashed near Canterbury. Five of the crew of JA931 were British and the other two were from Australia and Norway and all but one survived the war; this was air gunner, twenty-year-old Sergeant Bernard Stuart Lovell. He was killed on the night of 22 October 1943 while flying on another operation with 7 Squadron over Kassel.

No. 56 Squadron returned from Martlesham on 15 August and 197 Squadron, which had been standing in for the unit, flew back to Tangmere. In terms of what was important, the news of the death of thirty-one-year-old Flying Officer Jean de Selys Longchamps was what everyone was talking about. de Selys Longchamps had been flying Typhoon EJ950 on a Rhubarb sortie to Ostend and was landing at Manston when the aircraft suddenly crashed to the ground. It was claimed that de Selys Longchamps' Typhoon had been hit by flak as it had flown over the coast of Belgium and it had been badly damaged. There were others that saw the aircraft crash, who thought that the tail of his Typhoon had just fallen off and that de Selys Longchamps had become another victim of the aircraft's structural failures.

The last mention of de Selys Longchamps in the Manston ORB before he was killed was on the 13th. It stated: 'In the afternoon Flying Officer de Selys and Pilot Officer Callatay tried their luck but turned back from the coast at Dunkirk on finding only thin cloud at 2,500 ft.'. The ill-fated sortie on the 15th is only mentioned after reporting that de Selys Longchamps had been killed. It stated that of two other pilots from 3 Squadron, Flying Officer Shwarz had attacked a railway bridge at Amien and Flight Lieutenant Collins had bombed an airfield. There is no mention of what de Selys Longchamps did on this raid and it is possible that he had returned early to Manston because of whatever problem caused him to crash.

Flying Officer de Selys Longchamps was buried in Minster Cemetery with full military honours. Although he was gone, he was not forgotten; on Friday 16 August 2013, a memorial service was held in the cemetery to mark the seventieth anniversary of his death. The service, which was organised by Wings of Memory, was attended by officials from both the RAF and Belgian forces, with General van Calenberge and the Lord Lieutenant of Kent attending. Members from the Royal British Legion and Royal Air Forces Association were represented, in addition to cadets from Nos 3 and 609 Squadron Air Training Corps.

There was also flypast by a Typhoon (Eurofighter) of 3 Squadron and two F-16s of the Belgian Air Force, but unfortunately a flypast by a Hurricane from the Battle of Britain Memorial Flight had to be cancelled because of the weather. Wreaths were lain on behalf of the Belgian Air Force, the Belgian Embassy, the Royal British Legion, and many other organisations. After the service, everyone retired to the British Legion Club in the local village of Minster, where a wonderful spread had been laid on and there was an opportunity for veterans and others who had attended to exchange stories.

During August, the 4th Strategic Air Depot of the United States Army Air Force moved into Manston with field mobile units under the command of Master Sergeant Joen R. Campbell. The role of the unit was to make on-the-spot repairs to P-47, P-38s, and P-51s that had landed at Manston with battle damage and were in the need of being put back into service.

American involvement at Manston was urgently needed and, on 17 August, a badly damaged B-17, 42-5712 *My Prayer*, had landed on the grass after taking part in a raid on Schweinfurt, with just the pilot and two crew on board. Seven other members of the crew had baled out 15 miles south-east of Frankfurt after it had been attacked by fighters. The pilot was 2Lt James D. Judy, the co-pilot was 2Lt Roger W. Laya, and the only other man on board was engineer Sergeant Cherry, who had been badly burnt after fighting dozens of fires, but had prevented the aircraft from burning up in the air. The aircraft was from the 322nd Bomb Squadron, 91st Bomb Group, and had taken off from Bassingbourne.

No. 198 Squadron arrived at Manston with its Typhoons from Bradwell Bay on the 23rd under the command of Squadron Leader J. Manak. The Manston ORB wished the squadron well, but, in a matter of days, the unit would suffer several major setbacks. 'The swings and roundabouts' of Fighter Command continued to turn and, on the same day, 56 Squadron returned to Martlesham.

The 27th was a very busy day at Manston. Twelve aircraft were sent out to attack enemy tanks that had been reported to be in an area to the west of Ostend. The force included four Typhoons each from 3 and 198 Squadrons and four Hurricanes from 164 Squadron, but despite an extensive search, no tanks were found. The Commanding Officer of 3 Squadron was Squadron Leader S. R. Thomas, DFC, AFC, who had taken over from Belgian pilot Squadron Leader de Soomer just two days before.

Later on, a large number of B-17 Fortresses and their fighter escort flew over Manston after a raid on objectives in north-west France and a number of them found it necessary to land. Fighters from 303, 316 (Polish) Squadrons, 341 (Fighting French), 403 (Canadian), 485 (New Zealand), and 504 Squadrons were among those that landed to refuel and rearm. One of the Spitfires from 403 Squadron, flown by Flying Officer Foster, ran out of fuel mid-Channel and glided back to Manston, only narrowly failing to make the airfield, but he avoided serious injury. A single B-17 also made a crash-landing at Manston, flying on just a single engine after being badly damaged by flak.

Five days after arriving at Manston, Squadron Leader J. Manak, the CO of 198 Squadron, was shot down while flying Typhoon JP613 on a Ramrod sortie over Holland. Manak's aircraft was seen to be hit by flak and trailing a white stream of glycol before hitting the sea and sinking 3miles off the Dutch coast. Although nobody saw Manak escape his stricken aircraft, he did manage to bale out and was rescued to become a POW.

With the loss of Manak, it has been claimed that the command of 198 Squadron was taken over by Squadron Leader C. C. F. Cooper, but there is no mention of him in the Manston ORB. On 31 August, it states that Squadron Leader Mike Bryan had assumed command and that Flight Lieutenant Smith had been appointed as a Flight Commander.

No. 841 Squadron lost an Albacore at the end of August, when Sub-Lts Rutherford and Hitch were flying above the Channel on patrol and the aircraft's engine cut out. They tried to get as far as Hawkinge, but had to make a forced-landing in a field and could not prevent the aircraft from running into a fence, where it burst into flames and burnt out. Fortunately, both officers escaped unscathed.

On the last day of August 1943, it was mentioned in the ORB that the renovation of the north-east to south-west runway was going well, with the work being carried out by bulldozers and scrapers of the 23 Airfield Construction Group. It was also noted that from 500 feet it looked fully equal to the description recently given to it by Squadron Leader Rose—'A pig's breakfast'. Within a matter of a few days, however, Wing Commander Raphael was happy to report that the 'pigs' breakfast' had been resolved and Manston had become an orderly airfield once more.

On the first day of September, nine Typhoons of 3 Squadron, led by Squadron Leader Thomas, were tasked with covering the withdrawal of a number of Bostons that had attacked Roosendall. As there were no enemy aircraft around, they decided to attack shipping in the Channel and shot up a large sea-going barge and another large motor barge. At dusk, four Typhoons of 3 Squadron and six from 198 Squadron attacked shipping off the Dutch Islands, severely damaging a flak ship and a number of other vessels, including an armed trawler.

No. 1401 Met Flight arrived at Manston on 3 September under the command of Flying Officer Balmforth. They were equipped with the Spitfire Mk IX, after being formed out of the 'PRATA' element of 521 Squadron. PRATA was the code name for early morning photo-reconnaissance flights made at the highest possible altitude. The role of the unit was to carry out reconnaissance flights over Europe in the build-up to D-Day and the unit was destined to remain at Manston until beginning of 1944. Its operations were rarely mentioned in the Manston ORB.

It was a sad day for 3 Squadron, when, on 5 September, Squadron Leader S. R. Thomas, who had only so recently taken over command of the unit, was lost after his Typhoon was hit by a piece of debris from a vessel that he had attacked and scored a direct hit upon. His aircraft crashed on the Dutch Island of Schouwen and Thomas was spotted running away from his burning aircraft, but there was no immediate news of his condition. In his absence, his deputy, Squadron Leader Ronald Hawkins, MC, AFC, took over command of 3 Squadron.

No. 198 Squadron also lost a pilot on the same day, when twenty-one year-old Canadian Flying Officer James Lloyd Darby was killed while flying a Tiger Moth, DE765, over the sea off the coast of Margate. The aircraft also belonged to 198 Squadron and it was probably being used as the 'Squadron hack' for communication duties and ferrying its pilots around. It was assumed that the aircraft was lost because of engine failure and not enemy action.

Despite all the offensive sorties being flown from Manston, it continued to be used regularly by aircraft from Bomber Command. On 6 September, four Halifaxes and a Lancaster that had flown on operations over Munich landed. American activity was becoming more frequent and the whole of 334th Squadron of USAAC, equipped with the P-47 Thunderbolt, also landed at Manston.

The following day, 263 Squadron arrived from Warmwell on a detachment, although the author of the Manston ORB was under the impression that the unit had been posted to Manston. The Commanding Officer of 263 Squadron was Squadron Leader Baker, DFC, and it had been only the second unit to be equipped with the Westland Whirlwind. As 137 Squadron had re-equipped with the Hurricane Mk IV, 263 Squadron was by then the only unit operating the Whirlwind, although it would soon re-equip with Typhoon. While at Manston, 263's 'Whirlibombers' took part in just two operations, against what was described as 'beach guns' at Hardlelot on the 8th. Seven aircraft bombed successfully, but five returned to Manston with what was described as technical trouble. The following day, they again attacked the same gun emplacements at Hardlelot, but returned to Warmwell the following day.

The weather on the 11th was so bad that there was no flying within a period of twenty-four hours, with the exception of one brave soul, described as 'an adventurous tiger'. He was a pilot from 609 Squadron, flying a Typhoon, who got airborne, but almost immediately decided to return and only found his way back after a great deal of effort flying below the cloud base.

What was described as the Spitfire 'XII' Wing arrived at Manston on the 15th. Made up of the Spitfires from 41 and 91 Squadrons, it was under the command of Wing Commander Harries, DFC. Both units, equipped with the Spitfire Mk XII, were based at Westhampnett in Sussex and had been detached to Manston to provide cover for seventy-two B-26 Martin Marauder medium-bombers that were to bomb objectives over northern France. Unfortunately, only the Spitfires arrived over the rendezvous point at North Foreland and the operation was called off.

The following day, Flight Lieutenant Sinclair of 3 Squadron flew an uneventful patrol above Dieppe, Peronne, and Boulogne, but saved all the excitement for his landing at Manston. A bit of a mix up on the airfield meant that his Typhoon collided with a Mosquito IV from 139 Squadron, based at Wyton, that had just made a belly-landing after a sortie to Berlin. On the way back, it had been attacked by an Fw 190 and had flown most of the way on one engine, so the runway at Manston was a welcome sight for the pilot. Despite the fact that both the Mosquito and the Typhoon were complete write-offs, nobody was seriously injured.

Manston was visited by Group Officer (Group Captain) Lady Mary Walsh on the 20th. She, along with resident Flight Officer (Flight Lieutenant) Pam Barton, inspected the WAAF contingent at Manston. The following year, in August 1944, Lady Walsh was to achieve the highest rank in the Women's Auxiliary Air Force, when she was promoted to Air Chief Commandant.

The Marauders of the USAAF were to become regular visitors at Manston and seventy-two of them arrived at Manston on the 21st, along with the Digby Wing that comprised of two Canadian units, 401 Canadian (Ram) and 416 Squadron (City of Ottawa). Their assigned objective was to bomb the airfield at Beauvais, but, as the force approached the target area, a strange site was observed. A number of parachutes were seen descending from what was recognised as a German Ju 52 twin-engined transport aircraft. As far as could be observed, the aircraft had not been threatened or attacked in any way and the only conclusion made was that the Ju 52's crew, having seen the fighters, did not want to risk 'mixing it' with 121 Spitfires.

On the same day, Flight Lieutenant Keen of the Intelligence Branch, who had interrogated the pilot of the Fw 190 that had landed at Manston in May, gave a talk to a number of pilots. He told them about methods of interrogation and tactics used by Air Ministry Officers to persuade German POWs to talk—it was noted that it was of great interest to all those who heard it.

The Hurricanes of 164 Squadron departed Manston to move to Fairlop in Essex on the 22nd, but, with the exception of a Channel patrol flown by an Albacore of 841 Squadron, there were few other movements. This was to prove largely uneventful, although Lieutenant Fisher and Sub-Lieutenant Morris did find a pilot who had ditched 33 miles off Foreness and were able to guide an air-sea rescue launch to pick him up.

A service was held in Canterbury Cathedral on 26 September to commemorate those who were killed during the Battle of Britain. It was attended by Wing Commander Raphael and twenty-five other pilots from Manston. The sermon was preached by the station padre, Squadron Leader Pallet, and later a similar service was held at Broadstairs. At Margate, the Under Secretary of State for Air, Captain Harold Balfour, took the salute, along with Wing Commander Raphael and a number of other officers from Manston.

Lieutenant-Commander Walsh of 841 (FAA) Squadron was flying an anti-shipping patrol on the 27th, when he decided to try a new way of luring enemy shipping out into the Channel. Having flown numerous uneventful patrols, he was a bit disappointed and so he dropped what he described as 'decoy lamps' into the water, in order to attract the attention of enemy vessels that might be in the area. Unfortunately, he attracted the wrong kind of attention and the full weight of the Luftwaffe fell upon him from above and only by some very skilful flying at sea level did he manage to get back to Manston safely.

In the early hours of 28 September, a Halifax from 10 Squadron, based at Melbourne in Yorkshire, landed. One of the crew, Australian rear gunner Flt Lt Girardau, gave a personal account of what caused them to divert to Manston, and it provides and interesting tale, probably quite typical of many other bomber crews that found themselves at Manston. Having taken off from Melbourne at 8.45 p.m. in Halifax *M for Mother* the previous night, he and his crew were returning from a sortie to Hannover when they realised they were off track. They found themselves over the city of Frankfurt and had been caught by the master beam of the searchlights, which coned them for eighteen minutes, during which time they were continually battered by flak.

The Halifax had been at 17,000 feet when it had first been caught by the searchlights, but, after taking evasive action and corkscrewing to try and escape, it was at just 3,400 feet. The damage to the aircraft was extensive, with a petrol tank holed, lights, radio, W/T, DR compass, rev counters, and petrol gauges all unserviceable. The bomb aimer had mistaken the pilot's order, 'prepare to abandon aircraft', and baled out, only to become a POW. Having climbed to 9,000 feet and with the navigator taking Astro shots, the Halifax was attacked just before the French coast by an Me 210; fortunately, the aircraft had entered 10/10ths cloud because none of the aircraft's guns were working. At 2,000 feet and with the flight engineer being uncertain about how much fuel was left, the crew were told to prepare to ditch in the Channel. At that point, there was fresh hope when the pilot spotted a

Pundit beacon, which he thought might be on the clifftops at Dover, so he flew directly towards it. As the aircraft got nearer, Girardau spotted what he recognised as Drem lighting on an airfield, and so he ordered that the colours of the day to be fired off. The Halifax headed into the approach funnel and Girardau saw the flare path. However, at that point, he still thought that it looked like a 'Q' site and so he shouted over the intercom, 'Hang on boys it looks like a Dummy drome!'

It was not a dummy airfield, of course, and when the 'chiefie' on the ground opened the exit hatch, the crew were told that they had landed at Manston. It was 4.35 a.m. on 28 September 1943, and Girardau's doubts about the runway had been due to the runway's wire matting type. An inspection of the aircraft astounded many of the ground crew because the flak damage was extensive and there were large holes near each of the crew's positions, but nobody had been injured. Flight Lieutenant Girardau later wrote: 'You will understand for our crew to have come across Manston when we did, was an event that assisted us to survive a tour of operations'. As far as he was aware, the Halifax was never returned to 10 Squadron and the bomb aimer, who baled out, survived the war as a POW.

15

The Fiftieth Month of the War

The ORB for RAF Manston during the period of late September and early October 1943 contains only a small number of entries directly concerning events on the station. A large number of entries relate to the activities of the Typhoon squadrons and their operations over Holland and northern France. Most of those will be of little interest to those purely interested in events and activities on the station at Manston.

A B-17 (42-3082 named *Double Trouble*) from the 333rd Bomb Squadron, 94th Bomb Group, crash-landed in the sea near Margate on 4 October after attacking German airfields along the French coast. All the crew were safe, but one them, Staff Sergeant Van Hooser, was seriously injured.

There were further losses for 3 Squadron on 5 October after eight Typhoons (Bombphoons) of 3 Squadron and eight from 198 Squadron attacked the Sinclair Petroleum Refinery near Ghent. Two aircraft and pilots were lost, including the Commanding Officer of 3 Squadron, Squadron Leader Hawkins, along with Warrant Officer la Rocque. They added to the loss of Flying Officer Foster, who had been shot down earlier in the day. Hawkins was the second CO of 3 Squadron to be lost in a month and the role of Commanding Officer was taken over by Squadron Leader A. C. Dredge, DFC, AFC.

There was another change of office on the same day, when Squadron Leader McIvor, the Station Administrative Officer, was posted out to 11 Group headquarters at Bentley Priory. The author of the ORB stated that he had devoted twenty-one months of untiring service in the interest of the station and wished him well in his new post, while expressing a cordial welcome to his replacement, Squadron Leader Wise.

A large convoy of vehicles arrived at Manston on 7 October, made up of thirteen 3-ton lorries, four Jeeps, and twelve motorcycles, transporting members of the RAF's 2750 Regiment, which were to take up quarters over the winter period. The unit was the first of three Territorial Air Force squadrons to be based at Manston, composite groups of RAF Regiment and

AA personnel. No. 2750 Regiment had been formed at Duxford in March 1943 and, prior being posted to Manston, it had served as an anti-aircraft unit at 122 Wing Airfield. The Squadron was made up of eight officers, thirty senior NCOs (CSs), and eleven other ranks.

The Commanding Officer of RAF Manston, Wing Commander Raphael, left the station on 10 October to attend a combined operations course at Larges, Ayrshire, in Scotland. While he was away, Squadron Leader John Michael Bryan, the CO of 198 Squadron, was appointed as his deputy.

The following day, two more RAF Regiment (composite units) arrived at Manston; No. 2814 Squadron led the way, with seven officers, eleven senior NCOs, and 139 other ranks. No. 2701 Squadron arrived a few hours later and it comprised of seven officers, forty-two senior NCOs, and 123 other ranks. They dug themselves in around the airfield and assumed their duties straight away.

In his role as the Deputy Commanding Officer of RAF Manston, Squadron Leader Bryan gave a talk to an appreciative audience in Broadstairs at the Wings for Victory Commemoration Ceremony. The highlight of the meeting was when the chairman of the Savings Committee made a presentation to Squadron Leader Bryan of a plaque and logbook.

There was some excitement among the local population during the night of 15 October, when, at 11.15 p.m., a Ju 188 was shot down and crashed at Brooks End, Birchington, where Wing Commander Guy Gibson had made his forced-landing in April. The Ju 188 was a modified version of the Ju 88 that had first flown in 1942, designed with pointed wings, square-cut vertical tail, and BMW 801 engines. This particular aircraft had belonged to 1/KG 6 and it had been attacked by a Mosquito of 85 Squadron that was based at West Malling. Its crew consisted of Flying Officer H. B. Thomas (pilot) and Warrant Officer C. B. Hamilton (navigator). No. 85 Squadron was the unit that Manston's CO, Wing Commander Raphael, had previously commanded, but he had handed over to renowned night-fighter pilot Wing Commander John Cunningham in January 1943.

The German pilot, *Leutnant* K. Geyer, managed to escape, although there is no mention of how he got out of his aircraft, but he was soon captured and taken as a POW, being kept on the station at Manston overnight until the intelligence officers arrived the following day. The navigator had baled out, but for some reason his parachute failed to open properly and he was killed. A search was carried out by the RAF Regiment, the local defences, and a Naval launch for other members of the crew, but they failed to find any trace of them. Three days later, during the early hours of the 18th, a body, which was later identified as the 'third' member of the Ju 188's crew, was found after it had been washed ashore on the coast near Birchington.

There were actually four members of the crew of the Ju 188, but it is unclear where or when the body of the missing fourth German airman was found—it was probably discovered in the wreckage of the aircraft. The three German airmen that were killed were *Obergefreiters* Otto Schmidt and Dietram Kretzschmar and *Feldwebel* Walter Flessner; they were all buried in Margate's St John's cemetery.

There was some success for 841 Squadron during the night of the 18th, when Sub-Lieutenant Davey and Sub-Lieutenant Young destroyed an enemy minesweeper after scoring two direct hits on the vessel. Another ASR patrol was carried out by the unit, but returned with nothing to report.

Group Captain Frederick Laurence Pearce visited Manston on the 27th, having only recently been appointed as the Senior Air Staff Officer of 16 Group (Coastal Command). The group captain had begun his military career as a driver for the Royal Artillery in 1918 before joining the Royal Flying Corps as an equipment officer and training as a pilot in 1920. Serving with 269 Squadron, he had been awarded the DFC for his action in June 1940 when he had led a raid against enemy warships at Trondheim.

The ORB mentioned the fact that, despite the all-enveloping 'Scotch mist' that had recently become a regular feature of the district, it did not deter the Group Captain from touring the station in the company of the station commander. He also inspected 415 Squadron, a Coastal Command unit, which was equipped with Albacores, on a detachment to Manston from Bircham Newton.

The last two days of October 1943 at Manston were fog bound, and the only movements were made by the Albacores of 415 and 841 Squadrons on the 31st. Two aircraft from 415 Squadron and five from 841 carried out anti-shipping patrols in the Channel, taking off between 7.10 p.m. and 5.45 a.m., but returning with nothing to report.

At the end of October, there is a strange entry in the ORB containing comments that were almost certainly made by the CO Wing Commander Raphael:

> Which brings us to the end of the 50th month of the war. Perhaps not such a lazy month as could be desired, due mainly of course to weather conditions; but 31 days, at my rate, during which we can feel that RAF Manston has managed to take a little 'disciplinary action' against one or two 'very poor types'.

What exactly Wing Commander Raphael meant by these remarks is not known, and we can conclude that they were his remarks because they were separated from other entries in the ORB by missing out a line. The Wing Commander was known for his strict attitude towards discipline and we can only assume that someone or some persons failed to match the CO's expectations, for whatever reasons.

On the night of 3–4 November, Manston was used again by aircraft from Bomber Command with two Wellingtons and a single Lancaster landing during the early hours. The Lancaster, from 106 Squadron, had been on a sortie to Düsseldorf and was running short of fuel. Its crew may well have been impressed by the size and comforts of a permanent station like Manston. They were based at Syerston, which had also recently undergone a transformation from a grass airfield to one with a 6,000-foot concrete runway. However, in a matter of days, the Squadron would be transferred to a newly built airfield, RAF Metheringham; this airfield was barely completed, with accommodation that was situated in the cold, bleak, isolated Lincolnshire countryside, with no heating and little in the way of running water—whether hot or cold.

The station commander, Wing Commander Raphael, threw a party on the 6th and the two most important guest were the Commander-in Chief of Coastal Command, Air Marshal Sir John Slessor, and the AOC of 11 Group, Air Vice-Marshal Saunders. Slessor was a forty-six-year-old veteran of the First World War, who had already served for twenty-eight years, having joined up in 1915 as an untrained pilot. He had learned to fly at Brooklands and had been awarded Royal Aero Club Certificate no. 1477 on 6 July 1915. By the end of the war, he had been Mentioned in Despatches on four occasions and appointed as the Acting Commandant of the Central Flying School. He had taken up the post of C-in-C Coastal Command in February 1943 and was awarded the substantive rank of air marshal in June.

It was noted that 1Lt George S. Jones of the USAAF, flying a B-26 Marauder, displayed brilliant flying skills on the 10th, when he managed to perform a wheels-up landing on just a single engine, with no injuries to his crew. The aircraft was part of the 554th Bomb Squadron, 386th Bomb Group, and was returning from an operation against German-occupied airfields in the area of Lille.

On 13 November 1943, twenty-six-year-old Flight Officer Pam Barton was killed in a flying accident at RAF Detling, when the aircraft (Tiger Moth EM902) in which she was a passenger collided with a fuel bowser while taking off in bad weather. Flight Officer Pamela Espeut Barton (2532), daughter of Henry Charles Johnson and Ethel Maude Barton of Barnes Surrey, was buried with full military honours on Tuesday 16 November in Margate's St John's cemetery, although the burial service was held in the station church. It was noted that, despite a bitterly cold wind, many people turned out, including most of the WAAFs under her command. A statement was made following her death:

> She will be greatly missed by everybody, but more so, it must be said, by the WAAF personnel who came under her command, who will remember her as a great officer, whose sympathetic nature and kindly disposition, coupled with her fine powers of organisation, did a very great deal to promote a happy spirit among the WAAF.

WAAF Flight Officer and professional golfer, Pamela Espeut Barton, who was killed in a flying accident at RAF Detling on 13 November 1942. She had over 600 WAAFs under her command at Manston.

It was later arranged for a trophy to be titled the 'Pam Barton Memorial Salver', which would be presented to the winner of the annual ladies amateur golf championship. The pilot of the aircraft, her friend Flight Lieutenant Angus Ruffhead of 184 Squadron, who may have been giving her flying instruction, survived the crash, but was killed on 6 January 1944 flying a Hurricane.

There were a number of high-ranking visitors to Manston towards the end of November, including Air Marshal Roderic Maxwell Hill, former AOC of 12 Group, and the Commander-in-Chief of the recently formed Air Defence of Great Britain (15 November 1943) on the 24th. After the formation of the 2nd Tactical Air Force on 1 June 1943, Fighter Command had been replaced by the ADGB, which had been specifically formed for the defence of British airspace. The ADGB organisation had first been established in 1925, but was replaced by Fighter Command in 1936, which it still was in all but name. At Manston, the C-in-C met Wing Commander Raphael and then met and held a conference with all the squadron and flight commanders to explain the new organisation of the ADGB.

At the end of November, as was his usual practice, Wing Commander Raphael summed up the month as far as he was concerned. His comments typically separated from the other entries in the ORB by the omission of a line:

Thus our record for November is completed. Although not without its dark days, taken from the operational point of view, November has been a busy and satisfying month, with the 198 show on the 30th as a Grand Finale—Something ventured, something gained: And so onto December.

The Grand Finale referred to by Wing Commander Raphael was the action by pilots of 198 Squadron when they had destroyed four Fw 190s and a Ju 188. Those credited with the Fw 190s were the Commanding Officer, Squadron Leader Bryan, Flight Lieutenant Fittall, and Flying Officer Abbott, while Flying Officer Williams claimed the Ju 188. There was another Fw 190 claimed as damaged by Flying Officer McDonald, as well as a number of vessels that had been attacked and left in various states of destruction.

The sortie on 30 November was Squadron Leader Bryan's last before he handed over his command to the former CO of 609 Squadron, Squadron Leader John Robert Baldwin. Baldwin wasted no time in getting among the action and, during his first sortie after taking over, he destroyed a Fw 190 and went on to become the highest-scoring Typhoon pilot.

The first day of December 1943 was one of busiest and most chaotic that the station had ever experienced, with 125 aircraft landing in less than two hours and fifteen minutes. Among the first aircraft to arrive were the Spitfires of the Hornchurch Wing, made up of 129, 222, and 66 Squadrons. They were followed by the best part of 122 Wing, made up of 19, 32, and 65 Squadrons and then ten B-17 Flying Fortresses and a couple of Liberators. The first Liberator had had one of its wings almost completely severed and it crashed while on the approach to land, while the second one crashed on landing and burst into flames.

A B-24 (number 42-40793 and named *Blondes Away*) from the 565th Bomb Squadron, 389th Bomb Group, exploded at 800 feet above Manston; it is not clear whether this is the one in the ORB described as 'bursting into flames'. The aircraft was captained by 1Lt Jack M. Connors and it had been damaged by fighters while flying over its objective at Solingen. None of the crew had the chance to escape and all nine of them were killed instantly.

Aircraft continued to arrive into the early hours of the following morning, most suffering some sort of battle damage from enemy action; among them were four Lancasters, a Mosquito, and another B-17. One of the Lancasters, *Q for Queenie* from 426 Squadron, had been flying around the North Sea for over two hours and thirty minutes on two engines after being attacked by night fighters and losing all its instruments, electrics, and hydraulics. The aircraft, which was returning from a sortie to Berlin, made a bumpy landing and collided with and sheared the tail off one of the B-17s that had been parked too close to the runway. Regardless, the crew of *Q for Queenie* praised their pilot, whose skill and courage had got them home. Another Lancaster that landed after a long flight to Berlin was *F for Freddy* of 625 Squadron; it had been attacked by night fighters, but its gunners claimed to have destroyed at least one Ju 88.

One aircraft that failed to make it as far as Manston was B-17G 42-31243 of the 427th Bomb Squadron, 303rd Bomb Group, based at Molesworth in Huntingdonshire. The pilot, Lt Allan Eckart, was forced to ditch the aircraft in the relatively shallow water of Pegwell Bay, just a mile or so short off the airfield, after it had run out of fuel while returning from an operation at Leverkusen in Germany. Some of the crew were veterans and were approaching the end of their tour of duty, with Sergeant Michael Musache, the right waist gunner, having flown twenty-seven operations, while 2Lt George Arvanties, the navigator, was on his seventeenth.

Eckart and his nine-man crew were picked up by an ASR vessel and taken to Manston. The aircraft had only been delivered to the USAAC at Cheyenne on 8 October and assigned to the 303rd Bomb Group at the end of October; being relatively new, it had not yet been given a name. In 1999, there were attempts to recover the wreckage of the aircraft, during an excavation that author John Williams was involved in. Several pieces of the B-17 were recovered, but unfortunately the watch, lighter, and cigarettes that were left behind in the top turret by Sergeant Francis Neuner were never found.

Crew of B-17 42-31243, captained by Lieutenant Allan Eckart of 427th Bomb Group at Molesworth. The aircraft ditched in Pegwell Bay on 1 December 1943 after running out of fuel.

On 1 and 2 December, many of the 188 visiting airmen flew back to their own airfields. In total contrast, the 3rd was exceptionally quiet, with the exception of 198 Squadron carrying out a number of ASR sorties.

On the 4th, Squadron Leader J. B. Wray, the Commanding Officer of 137 Squadron, flew his final sortie, attacking an armed trawler and an R-boat before safely returning to Manston. Wray handed over his command to Squadron Leader J. R. Dennehey, DFC.

The Coltishall Wing, led by Wing Commander 'Laddy' Lucas, arrived at Manston on the 5th, comprising of 64 and 611 Squadrons equipped with the Spitfire Mk IX, landing at 12.40 p.m. Such was the urgency of their first operation that, after a 'lightning' briefing and the aircraft being refuelled, they were all airborne again within thirty minutes.

Advanced intelligence had been received that a 5,000-ton tanker was about to move up the Channel and 184 Squadron was ordered to institute standing Channel patrols from 6.32 p.m. The unit was based at Detling and was equipped with Hurricanes; they were using Manston as a forward base. The first pilot to arrive on the scene, Flight Lieutenant Holland, failed to see the tanker, but spotted a minesweeper off Dunkirk and had flown over it before he realised what it was. He had made a sharp turn to make an attack, but, despite a vector given to him by Swingate radar station at Dover, he failed to find the minesweeper again and presumed it may have taken shelter in the harbour at Dunkirk.

The 6th was another busy day; at 10.47 p.m., a Wellington Mk XIII that had been sent out on an anti-E-boat patrol (Patrol Deadly) landed, having been diverted from Docking in Norfolk. At 1.30 a.m., another Wellington was diverted to Manston after carrying out a later version of the same patrol. In its maritime role, the Mk XIII Vickers Wellington was used as a torpedo bomber that carried ASV radar, but retained its forward turret.

The Coltishall Wing was still at Manston on the 8th because bad weather had prevented them from returning to their base. The weather was so bad that there was no activity either by day or night. The following day, the Manston ORB notes that the Coltishall Wing was in danger of becoming a permanent unit at the station and that another operation, which they planned to do on the 10th, was also cancelled because of the weather—the Wing was able to return to Coltishall later in the day.

No. 195 Squadron arrived at Manston on the 11th with its Typhoons under the command of Squadron Leader Taylor, but the operation, which it was to carry out with 3 Squadron, was aborted because of bad weather. The unit returned to its base at Fairlop, but was back at Manston again the next day to carry out another operation with 3 and 198 Squadrons. This operation was also cancelled and the disappointed pilots of 195 Squadron returned to Fairlop.

WAAF Assistant Section Officer (Pilot Officer) Kaleva was posted at Manston from Andover on the 13th to take up duties as assistant adjutant, replacing Section Officer E. M. A. Hench, who was posted to Church Fenton. As was usual on these occasions, the author of the ORB extended a cordial welcome to her, at what was described as the 'Manstonian Fold'.

Losses at any time were hard to bear, but those that occurred just before Christmas were particularly felt; on the 20th, Flight Lieutenant C. Smith of 198 Squadron failed to return from a long-range operation to Deelen in Holland. Typhoons from 3 and 198 Squadrons had been supporting the withdrawal of a number of B-17s from a bombing raid in Germany. No enemy aircraft were reported, but Smith's Typhoon was last seen emitting blue smoke before crashing into the River Loeuwen—he was presumed to have been killed.

During the evening of the 21st, a concert was held at Manston. According to the ORB, it distinguished itself by a very fine first performance, despite some tearing of hair and hollow groaning by the entertainments officer. He had become upset when his one and only leading lady had been posted to cook for the officers' mess at 'RAF Puddleby-in-Slush' and his funny man had been detached on a course for bringing 'up unruly airmen' the way that they should be. The concert was a roaring success and the final comment in the ORB was that the author hoped that concerts would become regular events. It sounds like whoever wrote up the ORB was more than a little 'puddled'.

On Christmas Eve, the former AOC 11 Group (which had now been incorporated into the Air Defence of Great Britain), Air Vice-Marshal Hugh William Lansden Saunders, arrived at Manston on what was described in the ORB as a 'very official and very important visit'. After his inspection, the AOC authorised the immediate attachment of a works flight to carry out urgent repairs and work on the north-east to south-west runway. This had previously been carried out mainly by station personnel, but had proved to be a huge 'headache' and a drain on its resources.

There were no night operations because of the weather, but, as the ORB pointed out, the station was operational in another sense. Later on during the evening, the Christmas Eve party was held in the station cinema. Every section was represented in strength. As the evening wore on, the 'jollifications' became more evident and judging by Napoleon's maxim, a 'good soldier is a happy soldier', RAF Manston this night must have comprised of some excellent airmen.

According to the ORB, there were no operations on Christmas Day because of the weather, but it may well have been that hangovers were still being nursed and the station was getting ready to eat Christmas dinner. At 12 p.m., it was said that a great peace descended on the camp and all station personnel who were not on duty were to be found in the various messes, where excellent meals were served up. An attack was made upon seventy-six turkeys and twenty-three chickens that had been 'permanently grounded', led by the station admin officer, who served with 'great gusto'. He was backed up by 'squadrons' of station personnel, who, coming upon the tail of the 'enemy', gleefully pranged lashings of turkey, mince pies, and Christmas pudding on to waiting plates.

On Boxing Day, the SHQ Christmas party was held in the station cinema and it proved to be a fitting conclusion to the Christmas festivities. It was noted that great goodwill and countenance, diffused with the most benign of expressions, were the order of the day and a good time was had by all. In the middle of all the Christmas festivities, the fact that 3 Squadron had moved out to Swanton Morley in Norfolk on 28 December seems to have been forgotten.

Wing Commander Raphael made his usual comments in the ORB, giving his opinion on the operational performance of the station during the month of December:

> On looking back, it is noted that the month of December which entered (operationally speaking) as a veritable 'Lion of Vengeance', lost a fairish proportion of its bite and thrust after the first few days, due of course to a most uncooperative clerk of weather. Towards the end of the latter half of the month however, the 'Lion of Vengeance' appears to have been considerably revived by the efforts of No 3 Squadron (commanded by Squadron Leader Dredge), who, just before leaving for another station, achieved as a parting gift, first class results on all targets attacked.
>
> This twelfth month ended thus well and RAF Manston looks forward to playing well its allotted part in the year of Victory—1944.

A B-24 Liberator bomber, a type that was often seen at Manston in 1944. Along with the B-17 Flying Fortress, it was the workhorse of the USAAFs 8th Air Force.

16

The Year of Victory?

The year 1944 began with the news that Squadron Leader Baldwin, the Commanding Officer of 198 Squadron, had been awarded a Bar to his DFC, but the news did not prevent him taking part in an operation against a 6,000-ton blockade runner in the English Channel. The Typhoons of 609 Squadron joined those of 198 Squadron in attacking the vessel near Boulogne, with the former acting as 'spotters', which kept a lookout for enemy fighters, while the seven Typhoons of 198 flew through an intense curtain of flak. The vessel was hit by both cannon fire and rocket projectiles and was badly damaged below the waterline, while five of 198 Squadrons Typhoons were damaged by the flak, but managed to return to Manston.

Nos 198 and 609 Squadrons were active on both the 2nd and 3rd with excellent results, but the best day was on the 4th, when Squadron Leader Wells led 609 Squadron on a sweep of enemy airfields after a force of B-17s had been attacked by enemy aircraft operating out of Dutch airfields. At Gilze-Rijen, 609 Squadron destroyed seven enemy aircraft, including six Dornier 217s, originally designed as a bomber, although some had been converted for the night-fighting and intruder role. Two of them were destroyed on the ground, with Flight Lieutenant Davies and Pilot Officer Watts among those being credited, while one of them was credited to the 'squadron'. No. 198 Squadron also got in on the act and Sergeant Fraser destroyed another Do 217 to bring the total for the day to eight.

Major-General Miles from Eastern Command visited Manston on the 6th and, after he had been shown around the station, it was claimed that he was pleased with what he saw. The purpose of his visit was to inspect those units of the Army that were based at Manston, but to also improve RAF-Army relations.

Squadron Leader Mike Bryan returned to Manston on 13 January on a visit from 11 Group HQ; to 'keep his hand in', he took part in a Ranger operation with 198 Squadron and shot down an unidentified twin-engined training aircraft. During the same sortie, two Bf 109s were destroyed by Flight Lieutenant Niblett and Pilot Officer Lomas, but unfortunately the unit suffered the loss of Australian Pilot Officer Lamar, whose aircraft was hit by flak. During another sweep of airfields south of Paris later on, another three Ju 88s were destroyed by Flight Lieutenant Dial and Flying Officers Flamando and McDonald. Additionally, a second training aircraft, a single-engined type used by the Luftwaffe, was shot down by Warrant Officer Allan.

There is a mention in the ORB of the 'Manston Wing' for the first time on 15 January, when a long-range sweep by six Typhoons of 609 Squadron was cancelled because of the weather. Presumably, the Wing was made up of 198 and 609 Squadron's Typhoons, but there is nothing to confirm that in the records. At the same time, reports were received of a 5,000-ton vessel entering the harbour at Dunkirk and 415 Squadron carried out three anti-shipping patrols. These patrols were uneventful, except for the fact that the Albacore of Flying Officer Gates and Lieutenant Rice (USA) was fired upon by an unidentified aircraft.

During the 16th and 17th, weather was abysmal and there was very little operational activity; it was much the same on the 18th, except for the ADGB band visiting Manston. The band performed in the station cinema, where it was said that the bandsmen applied themselves to their instruments and played with a will, while their spirited rendering of hunting tunes as a grand finale must surely have induced every fox in the neighbourhood to take cover. It was said to have been a very pleasing performance, which was highly appreciated by the 'Manstonian' audience.

On the 19th, the Kentish version of the 'Scotch mist', as the author of the ORB described it, continued and there was no operational flying during the day. The bad weather continued into the night and the only movement was by a 415 Squadron Albacore that carried out a weather-reconnaissance flight over the Channel.

The following day, the weather had improved and the action continued after a report that a large German merchant ship had been hit by shore-based guns; nineteen Typhoons from 198 and 609 Squadrons took off under the leadership of Squadron Leader Wells. The vessel had been identified as the *Münsterland*, but Flying Officer Eagle of 198 Squadron confused the issue by saying that he had seen another German ship damaged near Calais. Additionally, other members of the squadron, who had previously carried out attacks upon the vessel, did not think it fitted the *Münsterland's* profile.

Two Typhoons from 609 Squadron, with Flying Officer Stark leading, went off on a reconnaissance sortie and filmed the vessel. When it was viewed, it was thought that the superstructure differed from that of the *Münsterland*. However, they were wrong and the vessel was, as originally identified, the 6,408-ton *Münsterland*, which had been regularly used by the Kriegsmarine as a blockade runner. It was understood that the vessel had run aground in fog west of Cape Blanc Nez, and had been targeted by the radar-assisted, long-range 14-inch guns based on the cliffs at Dover, named *Winnie* and *Pooh*.

At the same time as efforts were being made to identify the *Münsterland*, a section of two Typhoons from 198 Squadron took off on a weather-reconnaissance flight to St Omer. The section was led by Flight Lieutenant Curtiss at 800 feet. Close to St Omer, his Typhoon was seen to catch fire and crash into the sea, killing him. The body of twenty-two-year-old Richard Osborne Curtiss (63456), from Watcombe in Devon, was later recovered and buried in Le Parcq Churchyard.

During the night, 415 Squadron Albacores flew six air-sea rescue patrols and Squadron Leader Cowan and Flying Officer Wood were vectored on to an enemy vessel 5 miles off the coast of Calais. After carrying out two runs against what was later believed to have been an enemy destroyer, contact was lost with the Albacore crew and a search for them began.

A second Albacore took off, with Flying Officers Thommson and Bartlett observing and attacking the leader of two destroyers near Le Touquet with six 250-lb bombs, and a fire was seen to break out on the leading vessel. Flight Sergeants Perry and Jobe followed up this attack with another six 250-lb bombs, but the flak was very heavy and accurate, making it difficult to assess the damage to the lead destroyer, but it was claimed as Cat 'B'.

The following day (the 21st), the search continued for the crew of the 415 Squadron Albacore, and six Typhoons of 609 Squadron were sent to the coast of Calais where the flak was again very heavy and accurate. A Walrus air-sea rescue aircraft was seen circling a submerged white launch and dinghy near the Goodwin Sands. At the same time, reports were being received of a downed American bomber off the coast of Cape Gris Nez. Another dinghy was then spotted, which had been dropped by a Spitfire that then escorted the Walrus. Shells from German shore-based guns targeted the Walrus, but it was escorted safely to the shore by 609 Squadron. It is not mentioned who, if anybody, were actually rescued.

On the evening of the 23rd, a large number of officers from RAF Manston assembled at Doone House, Westgate, to welcome a guest for a party to celebrate recent events and air victories. Doone House had been a preparatory school before the war, but it was evacuated in 1940, moving to premises in Hertfordshire. Two of the VIP guests were Air Vice-Marshal Frank Linden Hopps, AOC of 16 Group (Reconnaissance Coastal Command), and Group Captain Frederick Laurence Pearce, the SASO of 16 Group.

Both officers were veterans of the First World War and Hopps had joined the Yorkshire Light Infantry in 1916 before re-mustering to the RFC in 1917. He had been the AOC of 16 Group since July 1943 and before that he had been the commanding officer of the 'Strike and Search' force in North Russia.

Pearce had been a driver in the Royal Artillery in 1917 and re-mustered to the RFC the following year. In July 1940, then a wing commander and CO of 269 Squadron, he had been awarded the DFC for leading an attack on an enemy airfield and scoring three direct hits on a cruiser during a raid on Trondheim in Norway.

The evening was said to have been a great success, apart from the fact that Wing Commander Raphael's 'heart was in his mouth' when there was a power failure just as things were getting under way. However, the night-visual capacity of all those present did a great deal to keep the party on course and make it a success.

Group Captain J. A. M. Faraday, the Deputy Provost Marshal, arrived at Manston on the 26th for a meeting with Wing Commander Raphael and it was noted that there was much shining of buttons and cleaning of webbing, especially among the ranks of the RAF Police. Unless one has been in the RAF, it is hard for anyone to understand the relationship between those ordinary airmen in the ranks and the RAF Police, which are known as 'snowdrops' because of the white band they wear around their hats. Typically, RAF Police through the ages have been known to spend a lot of their time chasing up airmen for haircuts and other minor breaches of regulations to file a Form 252 (Charge Form) against them. No love has ever been lost between them.

Flying Officer Custance was credited with introducing some exceptional talent to the RAF Manston cinema on Friday 28 January, which included artists such as John Clements and John Gielgud. Clements had appeared in films such as *Four Feathers* and, more recently, *Ships With Wings*, whereas Gielgud was more a stage star of the theatre and Broadway (although he had also appeared in films). It was claimed that rafters of the cinema had never rung with so much applause, and much appreciation was given to Eric Winston, his boys, and the officer who organised his visit, Fg Off. Custance.

Manston Wing's most successful day was noted as being 30 January, following 198 and 609 Squadrons fighter sweep between Rouen and La Roche. No. 198 Squadron, led by Squadron Leader Baldwin, had destroyed nine Fw 190s, with another as a probable and three as damaged. Among those credited with two 'kills' were the CO, Squadron Leader Baldwin, Flight Lieutenant Niblett, Flying Officer Williams, Warrant Officer Stanley, and Flight Sergeant Crouch with one and another as a probable, while Flight Sergeant Dall claimed one as damaged. Four of the Typhoons returned safely to Manston while another two landed at Frinton.

Seven Typhoons of 609 Squadron, led by Squadron Leader Wells, crossed the Channel and made land at Le Treport, attacking a number of enemy airfields with some success. Squadron Leader Wells and Flying Officer Shelton destroyed a Bf 110 on the ground at Roye Airfield, while Wells and Warrant

Officer Bucchanan went on to another airfield and damaged two Ju 88s. Flying Officer Detal destroyed another Ju 88 on the ground and then four Fw 190s were seen flying between Seine and Gisors. In the ensuing battle, Flying Officer Stark destroyed one of them and, during a prolonged dogfight, Flying Officer De Moulin destroyed another. All the aircraft returned to Manston, although one Typhoon from both 198 and 609 Squadrons was damaged. This action was noted as being the best ever performance carried out by the Typhoon aircraft.

After this operation, four Typhoons took off, led by Squadron Leader Baldwin, on a special sighting reconnaissance mission that had been ordered by the Air Staff to be carried out in the area of Ostend. Unfortunately, the mist and general weather conditions prevented the operation from taking place and Squadron Leader Baldwin aborted the operation.

Another two Typhoons of 609 Squadron took off on weather-reconnaissance sorties, while another two went off on shipping-reconnaissance sorties and spotted two coasters of 600–800 tons off the coast of Boulogne. The only other patrol carried out was by 415 Squadron, which sent a single Albacore on another weather-reconnaissance sortie.

At the end of January, as was his practice, Wing Commander Raphael made his usual remarks about how, in his opinion, RAF Manston had performed. He does not name the newly appointed public relations officer, but it was in fact Flight Lieutenant Corcoran:

> The Manston success for the month of January need not be reiterated here, and our Public Relation Officer so far, has had no trouble justifying his existence. The sincere hope now expressed is that the Manston Wing will continue to give the Hun an ever increasing number of 'headaches' such as have been handed out by the squadrons this month.

The first few days of February were again affected by the weather and only a small number of operations were carried out. On the 3rd, 98 and 609 Squadrons moved out to Coltishall at first light, in preparing for a long-range sweep of Dutch airfields in connection with a large-scale fortress raid on Wilhelmshaven. Owing to bad weather, the operation was cancelled and they returned to Manston during the afternoon. Meanwhile, the Coltishall Wing had moved to Manston to support a raid by Mitchells on objectives in northern France, but, as the operation was postponed, they spent the night at Manston.

The evening of the 5th was brightened by a performance by the Blackout's Concert Party; they were first-class Canadian artists, who, at the request of the Commanding Officer, had brought their show to Manston. Apparently their fame preceded them and the queue outside the cinema was long and dense, with the auditorium at full capacity. It was claimed that the show was even better than what everyone had been made to believe and, unfortunately, the remnants of the queue had to be propelled away from the cinema doors. The ORB expressed sincere thanks to the Blackout Concert Party for an 'evening of wizard entertainment' and to Wing Commander Raphael from organising it.

The following day, nine Typhoons of 609 Squadron left Manston to take part in an air firing course, but there is no mention of where it was to take place. Due to the weather closing in from the north of Manston, six Mosquitos from 464 Squadron and another five from 21 Squadron were diverted to Manston after carrying out operations in northern France.

On the 10th, 198 Squadron sent off two Typhoons, led by Warrant Officer Stanley, to investigate a burning wreckage that had been reported off the coast at Dungeness. In poor visibility, the two aircraft became separated and, after Warrant Officer Stanley had reported over the R/T that he was off the coast at Bolougne and had been hit by flak, they were ordered to land. One aircraft landed safely, but nothing further was heard from Warrant Officer Stanley and, despite a search by another four Typhoons, there was no sign of him or his aircraft.

On 14 February, a number of reunions were held, and some catching up was done when Squadron Leader Mike Bryan returned to Manston to meet up with fellow members of his former unit: 198 Squadron. In addition to this, 3 Squadron returned to Manston from Swanton Morley on what was described in the ORB as a seven week's sojourn in Norfolk. Wing Commander Rutting, the Commanding Officer of 415 Squadron, visited Manston to meet his officers and airmen and inform them about the unit's progress in working on the new equipment: the Vickers Wellington.

At 11 a.m. on 18 February, six Typhoons of 198 Squadron, led by Squadron Leader Dall, took off to escort a number of Mosquitos on what was described as a 'special mission' in the Pas de Calais area. It is not mentioned in the ORB what this mission was, but it is now known that it was the attack on the Amien Prison (Operation Jericho)—the mission was led by Group Captain Charles Pickard. Dall had had no time to brief his pilots and he was the only one who knew where they were going and what they were doing.

The purpose of the operation was to provide close support to the Mosquitos of 21, 464, and 487 Squadron, which were to make the attack on the walls of Amien Prison. The weather conditions were appalling and the formation of Typhoons was broken up in a snowstorm; three of the aircraft, flown by Flight Lieutenants Lallemant, Niblett, and Flying Officer Armstrong, returned to Manston. The other three Typhoons carried out the mission and later landed at Tangmere.

There was a raid by enemy aircraft in the vicinity of Manston on the same day of the Amien Prison operation. The raid was noted as being on a larger scale than others that had taken place recently. Despite the fact that enemy aircraft were overhead and could be seen quite clearly, the rules of engagement prevented the AA guns at Manston from opening fire. The rules stated that, unless a potential target was below 4,000 feet or unless it was illuminated and positively identified as hostile, the AA guns were not allowed open fire. This was because the canopy searchlights had recently been removed from Manston, despite the fact that Wing Commander Raphael had strongly protested about this decision. As a result, a number of low-flying enemy aircraft had recently appeared, seemingly deliberately 'stooging' over the airfield, aware that they would not be fired upon. Subsequently, the CO wrote a letter to 11 Group requesting that they be reinstated immediately and the response was that some searchlights would be provided by the 21st Army Group. It was noted in the ORB that '[they] now wait for low flying Huns'.

During the afternoon, 609 Squadron returned from its air firing course and the evening was very quiet, with nothing to report. A heavy raid by American Fortresses and Liberators caused 198 Squadron to be scrambled during the afternoon of the 20th to the area around Amien, but the weather conditions prevented it from getting involved in any action. A large number of the fighters and bombers involved in the operation landed at Manston and, at approximately 4 p.m., a Liberator exploded in the air close to the airfield while trying to make an emergency landing—all the crew were killed.

The Liberator was a B-24H (number 42-7529 named *Coral Princess*) and belonged to the 579th Bomber Squadron, 392nd Bomb Group, and it was being flown by Lieutenant B. Peyton. At the time of the incident, the Liberator was approaching the airfield to land. After the explosion, *Coral Princess* crashed into a field near the local village of Acol. The aircraft was based at Wendle in Norfolk and it was commanded by Colonel Irvine A. Rendle; *Coral Princess* had crashed while returning from its twentieth mission, which was to attack Halberstadt. Although the flak was considered to have been light, five aircraft had been hit and damaged by a combination of both flak and fighters.

Four of the crew—Lieutenant John B. Payton (the captain), 2Lt George Gregory, Sergeant Bermarr J. Burglund, and Sergeant Roy E. Welch—were buried in Cambridge cemetery. The bodies of the other members of the crew were repatriated for burial in the USA.

On the 22nd, 126 Airfield Wing (Canadian), 404 Squadron, and 412 Squadron—all under the command of Squadron Leader Keefer (appointed wing leader on this day)—landed at Manston when they were unable to make it back to their own airfield. Additionally, the airfield provided a safe haven for a large number of American fighters and the odd Spitfire that had taken part in a raid, supporting Flying Fortresses over France. During the day, 100 Spitfires landed at Manston, along with twenty-six Marauders, ten Mustangs, eight Thunderbolts, and five Lockheed Lightnings.

The nose artwork of B-24 Liberator *Coral Princess* that crashed in a field near Acol on 20 February 1944 with the loss of its crew.

Three important visitors arrived at Manston on the 26th, the first being General Glow, followed by the Director of Medical Services, Air Marshal Sir Harold Whitingham. The AM was a renowned authority in tropical medicine and had been awarded the Chadwick Gold Medal for his research into Sandfly fever in Malta between the years 1921 and 1923. The third distinguished visitor was Captain Harold Balfour, the Under Secretary of State for Air, who was meeting with the CO, Wing Commander Raphael.

A summary of the movements for the month of February 1944 stated that there had been 1,919 landings during the day and seventy-six at night, but total operational landings were 1,378 by day and thirty-six by night. Comments, possibly those of Wing Commander Raphael, stated that taking into consideration the time of the year it is satisfactory to note that the past month included remarkably few dud days due to adverse weather condition.

No. 3 Squadron departed Manston on 6 March for Bradwell Bay in Essex and it would later become one of the first units to counter the threat from the V-1 flying bomb. The unit had built-up close connections over the years, and it had been one of the first squadrons to arrive at Manston immediately after war was declared on 1 September 1939. It would return again in the future, but by that time the war had been won and many things had changed.

The Inspector General of the Women's Auxiliary Air Force, Group Officer (Group Captain) Beryl Constance Beecroft, visited Manston on 8 March, accompanied by Squadron Officer (Squadron Leader) Ramsden. Together they carried out what was described as an intensive inspection of the WAAF living quarters and the group officer announced that she was very pleased with what she had seen. She left the station saying that she hoped all the WAAF working at Manston realised what an important airfield they were working on.

There were no operations during the morning of the 11th, but, during the afternoon, a section of 609 Squadron was scrambled to search for a pilot that had been reported being seen in a dinghy off the coast of North Foreland. Owing to poor visibility, they failed to find anything, but Flying Officer Eagle of 198 Squadron, who was listening in on the vectors being given to the 609 Squadron pilots over the R/T, joined the search. He then spotted a patch of oil 7 or 8 miles from North Foreland before spotting the dinghy with the occupant waving a red flag. Eagle then flew towards two ASR launches that he had seen were stationary and guided them towards the dingy, making sure he had been picked up before leaving the scene. He later found out that the man in the dinghy was a Polish pilot, Flight Lieutenant Kiewice, from 315 (Polish) Squadron.

Brigadier B. R. Britten and Wing Commander Mayew arrived at Manston on the 16th to inspect the units of the RAF Regiment that were based there. It was said that the inspection went according to plan and everything was found to be in good order.

17

New Faces, New Runway

There were to be some major changes over the next few days, both to the composition of the Manston Wing and the station. On 16 March, 609 Squadron moved out to Tangmere after flying its final sorties from Manston the previous evening. The unit had strong connections with Manston and had first arrived in November 1942. The unit that replaced 609 Squadron was 197 Squadron, which had arrived at Manston on the 15th under the command of Squadron Leader Taylor. The unit had been formed at Turnhouse in November 1942, when it had been equipped with the Typhoon 1b.

On 17 March, 198 Squadron moved out of Manston for Tangmere and its operational status was taken over by 183 Squadron that had also moved in from Tangmere on the 15th—they were also equipped with the Typhoon. No. 198 Squadron had flown its final sortie from Manston on the 16th after it had been selected by 11 Group HQ for an attack on a particularly special target near Ostend. No. 183 Squadron was operational at Manston on the 18th, with a section flying hourly patrols of a convoy passing through the Channel. On the same day, Spitfire pilots of 412 (Canadian) Squadron were on an ASR patrol when they spotted a downed American B-25 Mitchell bomber and a dinghy in the water near the mouth of the Somme.

The 18th was a busy day, and, during the afternoon, four Typhoons of 197 Squadron were flying their first operation since arriving at Manston: Ranger sorties over Belgium. One of the four pilots that took part was lucky to survive and Flight Lieutenant Button flew so low that his starboard wing hit the wireless aerial of a ship that he was attacking. The formation of Typhoons returned to Manston and Button made a successful crash-landing, but his machine was written-off.

During the day, the operational landings by aircraft of the USAAF included four Liberators, one P-38 Lightning, five P-47 Thunderbolts, nine Mustangs, and a single Spitfire. That night, into the early hours of the 19th, Manston had a number of visitors from Bomber Command, including a Halifax from 158

Squadron and three Lancasters from 166, 103, and 9 Squadrons, which had bombed Frankfurt. They had been part of a force of 860 aircraft, comprised of 620 Lancasters, 209 Halifaxes, and seventeen Mosquitos, twenty-two of which (2.1 per cent of the force) had been lost. The Lancaster from 9 Squadron, *S for Sugar*, had been badly damaged by flak and two of the crew, the wireless operator and mid-upper gunner, had abandoned the aircraft before the pilot had regained control.

On the 24th, there were operational landings by three aircraft of the USAAF—a single P-47 Thunderbolt, a Liberator, and a Mustang—and three of the RAF—a Mustang, a Wellington, and a Spitfire. There was an official visit by Brigadier Pickering and Major Brierly, both from a sub-district of the Canterbury Home Guard, who were there on business concerning the RAF Regiment. That night, a Halifax of 158 Squadron and a Lancaster of 57 Squadron landed because they were short of fuel while returning from a raid on Berlin. The Lancaster had been attacked by fighters over Berlin and was badly damaged, while the Halifax had lost its hydraulics, causing its brakes to fail; unfortunately, the pilot could not prevent it running into a gun pit, where it was very badly damaged.

There was what was described as a sitting conference on the 29th in the station commander's office, including officers from ADGB, 11 Group (Fighter Command), and 16 Group (Coastal Command). The purpose of the conference was in connection with the proposed commitments of Manston's new role. The airfield was to be used more in a supportive role for emergency landings and refuelling and less active in the offensive role.

April was a very busy month and 137 Squadron arrived back at Manston from Lympne on the 1st, having exchanged their Hurricane IVs for the Typhoon 1B in June the previous year. Having left Manston in December 1943, there may have been some familiar faces among those who had been stationed there. No. 137 Squadron was under the command of Squadron Leader Gunnar Piltingsrud, who had just taken over from Squadron Leader Dennehey. Piltingsrud was Norwegian and had served as a Flight Commander in 56 Squadron, where he had been recognised a natural leader. On the same day, 183 and 197 Squadrons, having become 123 Airfield Wing, departed for Thorney Island.

A section of 197 Squadron stayed behind at Manston and were sent off to search for a missing B-17 at 12 p.m. that had been reported down in the sea off the coast of Nieuport. They were followed in the search by three sections of 137 Squadron, but, apart from a small number of coasters and trawlers that were entering the port of Ostend, there was no sign of the B-17. A Walrus from 277 Squadron also took part in the search, but it later landed at Manston with nothing to report.

That night, at 10.21 p.m., Warrant Officer Carter of 137 Squadron took off on an anti-shipping patrol in the Dunkirk–Flushing area; however, shortly after he had begun his search, Swingate's radar reported that his plot had become weak and then it suddenly disappeared altogether. At first it was thought to have happened because he was flying at sea level, but when they tried to contact him there continued to be no response. There was no further news of the aircraft, so, at first light on the 2nd, two sections of 137 Squadron took off to search for him in the Calais–Ostend area, but there was no sign of Warrant Officer Carter or his aircraft.

There was no operational activity on the 4th, but it was noted that there were many 'bodies operational' on the new runway, which had been cleared for use. The following day, operational activity was 'nil' again, but, at 2.45 p.m., the new 9,000-foot runway was officially opened. This was three weeks ahead of the original date of 1 May, set at the conference in June 1943. On the first day, operational landings on the new runway were made by just eleven Thunderbolts, three Mustangs, and a single Dakota.

What is not mentioned in any detail in the ORB is the FIDO system, the installation of which was delayed because the system fitted at Woodbridge was still under trials and it was hoped lessons learned there could be implemented at Manston. Additionally, the fact that Manston was only fog bound for an average of thirteen days a year delayed the decision of which system to fit. There was also concern about the location of the fuel storage tanks on the Ramsgate Road at Cliffsend and that they could easily become a target for hit and run raiders.

As one can imagine, such a system as FIDO required huge amounts of fuel, and to avoid clogging up the road with convoys of road tankers it was decided to transport it by rail. Subsequently, a siding was built in the local village of Minster, near the long-established railway station that could accommodate up to twenty-five rail tanks. However, this still left the problem of transporting the fuel for 1½ miles up to the storage tanks on the Ramsgate Road. It was not just a problem of distance because the storage tanks on the south-eastern side the airfield were on higher ground and the fuel had to be pumped up a gradient of 150 feet.

To pump the fuel up to the airfield, a pump house was built by the rail siding, into which the petrol was drawn from the rail tankers by a four-cylinder Morris commercial stationary engine and a Bradford pump. Two pump houses were built alongside the storage tanks on the airfield to draw the petrol up from the Minster sidings; each pump house contained six Sulzer pumps, powered by Ford V8 engines. The petrol was pumped through a 6-inch pipe buried below the ground and the pumps could deliver 500 gallons a minute. It was an incredible logistical operation.

What was described in the ORB as 'an unsatisfactory state of affairs' was caused by exceptionally bad weather conditions in the south-east of England. On the 6th, and for most of the 7th, there was no operational activity. The only exception was that the Mosquitos of 605 (County of Warwick) Squadron arrived at Manston, posted in from Bradwell Bay. During the night of the 7th, Flying Officer Williams of 605 Squadron, flying a Mosquito of 'B' Flight, became the first pilot to land an aircraft on the runway at night.

Things began to get busy on 8 April, when 132 and 134 Airfield Wings arrived, comprising of 66, 310, 312, 313, 331, and 332 Squadrons. The aircraft were quickly refuelled and pilots briefed to escort a number of Mitchells and Bostons that were attacking marshalling yards at Herson. After the operation, pilots were ordered to return to their own airfield.

The first emergency landing on the new runway happened during the afternoon of 8 April, when a Marauder that had been badly damaged by flak touched down. The co-pilot had been badly wounded in the arm and he may also have been the first casualty to be admitted to the sick quarters during this year. Two more Marauders arrived soon after and one of them landed on one engine, which was badly shot up, with the navigator dead and most of the crew wounded. Other operational landings included those by six Thunderbolts, which needed refuelling, and a Mustang with engine trouble.

There was also an official visit on the 8th by Australian-born Air Vice-Marshal Don Bennett, the AOC of 8 Group Bomber Command; he was the officer that had set up the Pathfinder Force in August 1942. Bennett was one of the RAF's most experienced airman, having flown Flying Boats, instructed at Calshot in air navigation, served in the Australian Air Force, and worked as a civilian pilot with Imperial Airways. It is not mentioned in the ORB why he was visiting Manston, but it was almost certainly to do with the new runway and the installation of FIDO.

Two Typhoons of 137 Squadron were scrambled during the morning of the 9th to escort a damaged B-17 back to Manston. The aircraft was reported as being 15 miles to the north-east, but no contact was made and the two Typhoons returned to base. The only operational landing during the day was a single Mosquito.

The following day, another two Typhoons of 137 Squadron were sent off on a weather reconnaissance flight and two more to search for a B-17 that was reported as being in distress. They were recalled 20 miles off the coast of Dover because of bad weather. Two more aircraft were scrambled to search for another aircraft in trouble off the coast of Dover, and although they failed to find it they did see oil and black streaks on the surface of the sea. This may have been the same aircraft that the first two Typhoons had been looking for; four MTBs and a Walrus joined in the search, but there is no mention in the ORB as to whether the aircraft was found.

Search and rescue was becoming an important role for 137 Squadron and two Typhoons took off to escort a Walrus on the 13th that had landed on the sea off the coast of Neiuport—they failed to make contact with it. They did,

Above: No. 137 Squadron aircrew posing with a Hawker Typhoon 1B, with which the unit was re-equipped in January 1944.

Left: No. 263 Squadron airmen resting at Manston during a short detachment in September 1944..

No. 137 Squadron ground crew working on a muddy airfield in October 1943.

however, see a pilot in a dinghy being picked up by two ASR launches 15 miles off the coast of Deal. Another Walrus landed at Manston, with the pilot of an American P-38 Lightning picked up 25 miles north-east of Manston—he was taken to sick quarters.

There was what was described as a large and stimulating party held in the officers' mess at Doone House, Westgate, on the 16th; it was attended by a number of RAF dignitaries, including the Under Secretary of State for Air, Sir Harold Balfour, and the AOC of the ADGB, Air Marshal Roderick Maxwell Hill. It was noted that, although those partaking in the operation maintained a discreet silence, it was understood from neutral sources that a 'very' good time was had by all.

On the night of the 18th, what was probably the first 'heavies'—two Halifaxes, one was a Mk II from 78 Squadron and the other a Mk III from 466 (Australian) Squadron—from Bomber Command landed on the new runway. Both aircraft had lost an engine before they reached their objective and had flown over 800 miles to reach the safety of Manston. They were part of a force of 171 aircraft that included 139 Halifaxes and twenty-four Lancasters, with fourteen (1.2 per cent of the force) aircraft lost on this night.

There were complaints on this night about 'Sundry bodies' persisting in using the new runway as a shortcut when late for urgent appointments; it was claimed that some of those apprehended had found that it was not 'heaven sent' after all. Such was their punishment that it was noted that they may have resolved to give the tempting tarmac a wide berth in future. There is no mention of what exactly what was meant by this, but it is quite likely that it referred to the fact that airmen were using the runway as a shortcut to visit the Prospect Inn on the south-western side of the airfield. The village of Minster was also just across the runway, where there were a number of inviting hostelries that may have attracted them.

Ten Grumann Avengers of 848 Squadron (FAA) arrived at Manston on the 20th for duties in the English Channel. The Naval unit had been formed as a torpedo bomber squadron in June 1943 and was attached to HMS *Formidable*.

During the evening, a Walrus of 278 Squadron located three men in the water, who turned out to be American airmen from a downed aircraft that had only recently been lost. They were discovered in the sea just 8 miles off the coast of Calais and, although the Walrus was unable to take them on board, it was able to guide a HSL to pick them up. Two Typhoons from 137 Squadron were scrambled to cover the operation, but they landed without encountering any opposition.

That night, 605 Squadron's Mosquitos were out in force and flew a total of thirteen operational Intruder sorties and patrols, including one training sortie to Deelen. Among the targets were objectives in Bonn, the Rhine, and St Trond. One unidentified enemy aircraft and a Do 17 were damaged, but one crew, comprising Flight Lieutenant Hilland and Flying Officer Wilkinson, failed to return from their sortie to Hansdorf.

During the early afternoon of the 21st, Manston was warned by 11 Group that there three Marauders to the south of the airfield, whose crew were in trouble; however, the three aircraft subsequently landed. The first one, from the 572nd Bombardment Squadron at Matching, landed at 12.23 p.m. and was short of fuel, but the crew were alright. The second one, from the 496th Squadron, based at Stanstead, landed at 12.34 p.m. and it was badly shot up with the its fuel tanks damaged, one engine partly unserviceable, with the rear gunner having been killed. The third and final Marauder, from the 576th Squadron, landed with a full bombload, but had an engine out and had lost all its hydraulics. Its pilot performed what was described a 'very tricky' landing and the crew escaped unharmed.

There was a collision on the area of the new loop at the extreme western side of the runway between a work's lorry and an American Thunderbolt; as a result of this collision, it was noted that the lorry became 'airborne'. Subsequently, all units were instructed to inform their pilots to land on the northern side of the runway while workmen were operational on the southern side.

It was claimed that Manston was becoming a 'pranger's paradise' on 22 April after a large number of damaged aircraft had sought sanctuary there, starting off with the Marauder of the 574th Bombardment Squadron that landed at 11.50 a.m. with an engine out, but its crew safe. At 12.21 p.m., another Marauder arrived, diverted from Chipping Ongar, and it also had an engine out, overshooting the runway on its first approach, then landed safely. Shortly after that, a Spitfire landed with the burst tyre and there was no contact with the pilot because his R/T had been damaged. Then two Albacores from 415 Squadron landed and both of those had burst tyres.

Next came two P-47 Thunderbolts, one that had lost its hydraulics and had problems with is brakes, and one that had lost part of its wing after colliding with a train during a low-level sortie. Following this, five B-17s dropped in at Manston, each one bearing various states of battle damage; the first one had lost an engine, the second had lost two engines and was severely damaged by flak, and the third B-17 had one engine on fire. Although the fourth had two engines, it had been severely damaged by flak and the pilot was unable to use the flaps. The fifth and final B-17 had been extensively damaged by flak, but, despite all the damage to the Fortresses, none of the crew of the five aircraft suffered any wounds or injuries. The final landing of the night was made by a B-24 Liberator, which had lost two engines and also had problems with the hydraulics so the pilot was unable to use the flaps.

The 22nd was noted as being an exceptionally hectic day, with aircraft ending up in the most unaccountable and awkward places on the airfield. It was claimed that had it not been for the prompt and skilled actions of the crash parties and servicing crews, the airfield could have been easily mistaken for a very large and dispersed salvage dump. Operational landings were made by two Typhoons and fifty-two Thunderbolts.

That night, Squadron Leader Davies and Flying Officer Thompson of 415 Squadron, flying an Albacore, were involved in the search and rescue of the crew of a Liberator that had been reported down in the sea 60 miles east of Manston. Contact was made with the ASR launch and sixteen flares were dropped to illuminate the area, but there was no sign of the Liberator or its crew. The Albacore returned to Manston, but was soon sent out again; on its second sortie, the crew spotted a Walrus some 60 miles north-east of base, being towed by two ASR launches. It was unable to fly because of the weight and number of those on board, so for two hours Davies and Thompson orbited the Walrus and escorted it back to shore, which was safely reached at dawn.

There were what was described as three 'crash landings' on the 23rd, made by an Albacore, a Typhoon, and a Spitfire that had had both tyres blown out. There was also another incident on the runway when a Swordfish from 819 (FAA) Squadron taxied into a lorry, but fortunately nobody was injured. The FAA unit was on a detachment to Manston and there is no mention in the ORB as to what date it had arrived.

That evening, three Albacores from 415 Squadron took off on anti-shipping patrols in the Channel, and two of the crews claimed to have damaged a small number of trawlers that were being used for minesweeping. After patrolling for two hours, the crew of the third Albacore was recalled and ordered to jettison its bombload, but then it was almost immediately ordered to attack a lone E-boat that was described as a 'sitting target'. Unfortunately, just seconds before the second order was received, the bombs had been jettisoned and the crew were powerless to do anything about the enemy vessel. One can only imagine how frustrated the crew of the Albacore must have been and there may well have been a heated phone call between Manston and Swingate radar station after it had landed.

Four Halifaxes and two Lancasters landed on the night of the 24th after being damaged by flak and running short of fuel. One of the Lancasters caught fire in the area of flying control. At 4.30 a.m., an unidentified aircraft, presumed to have been an enemy machine, flew low downwind along the runway, firing its guns as it went. Some of the rounds ricocheted off the runway controller's cabin, but fortunately nobody was hurt; however, the identity of the aircraft remained a mystery.

Six Typhoons from 132 Squadron landed on the 26th to refuel before taking off on what was described as a 'special operation'. However, the day was described as being 'joyless', with two emergency landings made by Bostons. However, the aircraft removal party was not too bothered by the lack of action and one of them noted in the ORB that 'A wee restie willna do [them] any harrr-m'.

During the early hours of 27 April, it was reported that there were several enemy aircraft in the area again, described as playing 'tig' by flying in and out of the searchlights. It was noted that 'discretion played the better part of valour' and the enemy removed himself from the vicinity of the airfield after being challenged by the AA defences.

At 3 a.m., a Halifax from 424 (Canadian) Squadron landed with one of its port engines on fire after it had been hit by light flak while bombing Alnoyes railway sidings from 5,000 feet. The crew later stated that it was only them making an emergency landing at Manston, which prevented fire from reaching the fuel tanks and the aircraft exploding. One of the gunners had been wounded by the flak and he was taken to the station's sick quarters.

18

A Bad Day for Thanet

It was a bad day for the USAAF on 27 April; during the evening, two of its Liberators crashed locally within a few hundred yards of one another in a very short space of time. The first one to crash was 41-29509 from the 577th Bomb Group, 392nd Bomb Group, based at Wendling. It crashed at 8.10 p.m. into shallow water, approximately 100 yards offshore at Lodge Point, St Mildred's Bay, Westgate, killing five of the crew. The Margate lifeboat, *Lord Southborough*, was launched after receiving a call out at 8.11 p.m., but it could not render assistance because the remains of the aircraft were high and dry out of the water—help had to be given by the ambulance service and the military.

The aircraft, which had been named *Knuckle Head*, was under the command of Lieutenant Jacob Weinheimer and it was reported as having been hit by flak when in position 5055 N, 0320 E at 7.30 p.m. There were several dead and wounded on board the Liberator, which was being flown by second pilot 2Lt George C. Marshall, who was subsequently awarded the DFC for his efforts to save the aircraft. He was the only member of the crew that was not injured and he helped rescuers free his captain, Weinheimer, from behind the control column where he was trapped. He also rescued gunner Sgt Duffy out of the aircraft and dragged him onto the wing unconscious, before he was taken ashore on a stretcher.

Weinheimer's back and pelvis were broken, as well a number of his ribs, while tail gunner Sgt Robert Duffy had also broken his back. Second Lieutenant Gurwit, the navigator, had been injured, but his injuries were less severe than the others. All three were taken to Ramsgate Hospital to be assessed and, after their conditions had stabilised, they were transferred to the American 77th Evacuation Hospital.

The Liberator's port wing had been badly damaged by flak over Dunkirk and the ninth member of the crew, radio operator Sgt Parke V. Kent, had abandoned the aircraft, baling out over the town and becoming a POW. It was reported that Kent's parachute had been the only one on board the aircraft that had not been damaged.

After the war, Sgt Kent later gave his version of events leading up to the moments before he baled out and he stated that they had taken off from England after a very quick briefing at 2 p.m., but Kent claimed that only the pilot and navigator attended the briefing. The target was the marshalling yards at Châlons-sur-Marne (Châlons-en-Champagne). He was a spare radio operator and so he did not know any other members of the crew until that day. During the operation, they had bombed their objective before they were struck by flak near the French coast, causing the Liberator to enter into a slow spin. The engineer, S/Sgt Aughinbaugh, prepared to leave his position in the upper turret, followed by the co-pilot, 2Lt Marshall, who left his position and began to pull off his flak suit and oxygen mask.

Kent claimed that it had been his duty to clear the flight deck, open the flight deck door and bomb bays (which he said he did), and then await further orders. The aircraft was still descending, when the engineer appeared again and attempted to get his feet on the catwalk where Kent was standing. It was at that point that Kent claimed he was forced to bale out to make room for the engineer and the co-pilot, who had also left his position. Strangely enough, there is no mention of whether the co-pilot or 2Lt Weinheimer was at the controls or the time that he was wounded.

The Margate lifeboat had just returned to its slipway when its coxswain, Mr E. D. Parkes, and his crew saw a bomber flying low over the boathouse and, almost immediately, the crew caught sight of two parachutes. According to eyewitness accounts, the aircraft had already crossed the coastline and had flown very low over Margate at roof-top level, presumably to give the crew a chance of baling out.

At approximately 9.10 p.m., the second Liberator, serial number 41-29543, flown by Lieutenant Harold J. Larson, crashed into the cliff face at Cliftonville, killing eight of the crew. The aircraft was from the 706th Bomb Squadron, 446th Bomb Group (Bungay Buckeroos), based at Bungay. Having been badly damaged, it was flying on just two of its four engines and was heading for Manston. Its primary target had been V-1 sites in the Pas-de-Calais area.

The Margate lifeboat proceeded to the area where they had last seen the bomber, noted as being ¼ mile east and approximately ¼ mile off shore, but, in the meantime, it had crashed into the cliffs and exploded. On arriving at the scene, ropes were thrown to the two airmen that were seen to be thrashing about in the sea and the order was given for the boat to go astern, but almost immediately the engine stalled. Despite the anchor being dropped, the boat was nearly on the rocks. It was soon discovered that the problem was with the propeller of the lifeboat, which had snagged one of the two airmen's parachutes in water.

In the meantime, other boats had joined the rescue attempt and among them were two RAF high-speed launches—HSL127 and HSL169, from 27 Marine Craft Unit. Each of them pulled one of the American airmen from the water. The airmen were radio operator Technical Sergeant Carl M. Smith and top turret air gunner T/Sgt Edward Hilgeman. A couple of soldiers had also attempted to rescue the airmen and had swum out to sea and they were also picked up by HSL127.

The two airmen that had survived, Smith and Hilgeman, had been ordered to bale out just moments before the aircraft crashed and after being recovered from the water they were taken to Ramsgate Hospital. The lifeboat, the *Lord Southborough*, was not refloated until 1.30 a.m. and the propeller was not cleared until 3.30 a.m. on the morning of the 28th—the crew had a very long night.

There was an attempt by the *Thanet Gazette* to publish the story of the events of 27 April, but, as one can imagine, the censor at the Ministry of Information failed to pass it. The report gives only the bare facts about the incident, although it does provide an overall view of events:

> People of Margate are full of praise for the self-sacrifices of the pilots and crews of two United States bombers, who risked their lives in order that their aircraft should not crash on the town.
>
> Within an hour, two Liberators badly damaged in an evening raid over Germany flew at roof-top height over the town, unable to make a safe landing. The first made a belly-landing in shallow water near to the cliff, and the other, heading towards the sea, suddenly crashed with two engines on fire on the edge of the cliff at Foreness Bay. Unfortunately, a dozen of the men in the planes lost their lives.
>
> Four men in the first machine were found dead when rescuers, wading through the water, reached them. Four others were injured and taken to hospital. The tail gunner is unaware of what happened, for he was ordered by his pilot to bale out over enemy-occupied territory in consequence of damage to the aircraft, the tail nearly being severed from the main structure. One member of the crew is missing.
>
> It was getting dark when the second Liberator apprd. [*sic*] flying very low and obviously in difficulties. It had just skimmed the roofs of hotels and houses and seemed to be making for an open stretch on top of the cliff when two engines burst into flames. The next second it dived straight down and became enveloped in a mass of dense black smoke.
>
> Only two of the crew succeeded in jumping out just before the crash. They landed in the sea not far from the shore, and although soldiers waded out to try to reach them, they were quickly picked up by air-sea rescue boats which had been summoned to the first Liberator and were returning to their station. The Margate lifeboat was also launched.

The bomber blazed furiously while would-be rescuers stood by helpless. One wing and two of the engines fell off and fell on the promenade below. When firemen had overcome the flames, the charred remains of eight gallant Americans were recovered.

During the 1970s, the identity tags of Lieutenant George Hasner (the navigator aboard Larson's Liberator) were found on the beach. It is claimed that, after a storm that caused a lot of damage, it had moved a lot of sand and broke up a number of groynes. The ID tags had been found near the site of the where Liberator 41-29543 had crashed, along with small amounts of wreckage.

According to Mr Tony Ovendon, who was described as a 'beachcomber', in 1995 he had been put in touch with the person who had found the tags and borrowed them. He said the details on the tags were '0752488 T43 Mr G. P. Hasner' and gave his home address as '779 Lafond Avenue, St Paul, Minnesota'. Mr Ovendon then returned the ID tags to the man who had found them and suggested to him that they should be given to Hasner's family in the USA. As far as is known, the ID tags were not given back to the family and no attempt was ever made to contact them.

19

Towards D-Day

The ORB claimed that civilian workmen were treating the airfield like the 'acacia café', an institution that they could not be rid of at any price. Workmen were engaged in making alterations to the lighting system at both the east and west end of the new runway. The acacia café, described as a 'bone of contention' for service personnel, had just been declared '*out of bounds*' by the Americans to all American personnel.

As the first station personnel arrived in the morning, it was a regular feature for them to see a number of squadrons already on the airfield having arrived to refuel and rearm, with the petrol bowers merrily chugging along on their tracks. With the drivers looking over their shoulders while in a hurry to get to their action stations, it was sarcastically noted in the ORB that standing orders stated that they were to avoid running on to the grass at all cost.

As the first squadrons arrived in the morning, the order 'all guns-action-visitors' resounded across the tannoy, which was the signal for the RAF Regiment crews manning the AA guns to maintain an increased state of vigilance. That state of alert could last from half an hour to a half a day, until what was described as the last streak of silver had disappeared into the sky. At that point, the order 'all guns revert to normal state' was issued over the tannoy. The following are the figures for the number and type of aircraft that refuelled at Manston in April:

British		American	
Fighters	974	Fighters	351
Bombers	44	Bombers	88
Miscellaneous	111	Miscellaneous	45
Total	1,129	Total	484

This made a grand total of 1,613 aircraft that had landed at Manston due to a shortage of fuel and that may have been lost if the airfield had not been available for refuelling.

It was also quite normal to hear a message over the airfield tannoy at regular intervals, such as 'calling the aircraft removal part. Remove the Spitfire from the runway'. The only difference would have been in the type of aircraft to be removed or the number of them. There were fifty-six emergency landings by aircraft with battle damage or technical problems in April, with fifty-two of them landing on the airfield and four landing outside the boundary. The number of emergency landings in April was one Lancaster; eight B-17 Fortresses; five B-24 Liberators; three Halifaxes; thirteen Marauders; four Mosquitos; three P-38 Lightnings; one Mustang; and eight P-47 Thunderbolts.

RAF Manston had its own Servicing Wing in the form of a detachment from 86 Maintenance Unit, which was based at Sundridge Aerodrome near Sevenoaks. There was also much needed help and support from an American contingent led by Captain John Patrick English. The following is the number of aircraft that were repaired by the Servicing Wing in April:

British	American
Three Lancasters	Six B-17 Fortresses
Three Halifaxes	Two Liberators
One Mustang	Eleven Marauders
One Albacore	Three P-38 Lightnings
Five Typhoons	One Mustang
	Seven P-47 Thunderbolts

In the ORB, Wing Commander Raphael commented: 'Thus closes another month packed with interest and activity: but now we have the satisfaction of knowing that May will be a Busier busy month'.

May certainly started off as the wing commander had predicted. On the first day of the month, eighty Spitfires of 132 and 135 Air Wings landed to refuel prior to escorting a number of Bostons and Mitchells on a bombing operation. Operational landing during the day included those made by a Boston, a Liberator, and a Fortress. That night, the searchlight cones were exposed three times to help bomber crews in distress, but none of them landed at Manston.

The following day, beginning at 10 a.m., nine squadrons of Spitfires began to land for the pilots to be briefed and aircraft refuelled before they took off to escort a number of Bostons and Marauders to their targets in France. There was a three-hour postponement of the operation, during which time the pilots made excellent use of the rest and refreshment room that was provided for visiting aircrew.

At first light on 4 May, 137 Squadron despatched a single Typhoon on an anti-shipping patrol in the Channel, but the pilot, FS Eastbrook, failed to return. He was last seen attacking a trawler, firing at it with his aircraft's cannon, but then his aircraft was hit by flak and it dived into the sea.

That evening, it took flying control thirty minutes to 'home' a Mosquito that had lost the use of its port engine and had suffered severe damage to its rudder as a result of heavy and accurate flak. The aircraft had flown all the way from Cologne on one engine, but the pilot made a good landing and neither he nor his navigator were injured.

During the afternoon of the 6th, a contingent of WAAFs and RAF Regiment set out in 'battle array' to take part in the 'Salute the Soldier' parade in Broadstairs. It was noted that the parade was carried out with much usual verve and gained favourable comment, with the soldiers duly returning the salute.

On 8 May, Manston broke its own record for the number of daylight landings with a total of 357 being made in the day—116 of them occurring in just twenty-five minutes. Among them was a Mustang that landed with an engine on fire; as soon as that had been dealt with, there was a B-17 landing with its port inner engine on fire and both wings badly damaged by flak. Three of the crew were injured and were taken to the sick bay.

A Marauder with an engine out was next to land, but the most spectacular landing of all was made by a Liberator that had lost the use of all four of its engines. The aircraft had only one engine working when it began its approach, but that failed just before it touched down, leaving the Liberator gliding through the air. However, it made a good, but bumpy belly-landing and none of the crew were injured.

The incident with the Liberator may have been the occasion when Captain Bember, an American medical officer, arrived near the runway with his ambulance and staff, contravening all the traffic regulations on the airfield. An RAF officer, whose rank was obscured by his raincoat, told some of Bember's men to 'get the hell out of here', and Bember intervened and told him in no uncertain terms not to speak to his staff like that. The two men got into a heated argument before Bember realised that he was dealing with Manston's Commanding Officer, Wing Commander Raphael. They had a good laugh about it afterwards and neither bore any malice.

There were thirteen landings during the night of the 8th. Among them were those made by two Hawker Tempest V of 3 Squadron. Although 486 Squadron had been the first to receive the Tempest, it was 3 Squadron, based at Newchurch, which was the first to be fully equipped with the type. This is the first mention of the Tempest appearing at Manston.

There was a sad loss for 605 Squadron on the night of 10–11 May, when flying control spent some time attempting to 'home' a Mosquito, N5945, which was returning from a sortie to Venlo. The aircraft, whose crew consisted of Flt Lt Trevor McAlpine Laurence Woods and Fg Off. Kenneth Henry Ray (navigator), had been hit by flak over the Dutch coast at 12.07 a.m., and at one point they had considered ditching the aircraft near Walcheren. Attempts to 'home' the aircraft to Manston began at 1.30 a.m., but then nothing further was heard from them. At 2 a.m., North Weald informed Manston that the Mosquito had crashed near Dover.

There was a total of 291 landings on the 11th and there were also a number of interesting VIP guests visiting Manston, among them was General Goodrich of the United States Eighth Air Force and Air Chief Marshal Sholto Douglas, the Air Officer Commanding-in-Chief of Coastal Command. The other visitors were Air Vice-Marshal F. L. Hopps, the AOC 16 Group (Reconnaissance) Coastal Command and Major-General Llardet, Director of Ground Defence (RAF Regiment).

The next day, twelve B-17 Flying Fortresses landed at Manston, eight of them having been badly damaged on operations. These were followed by eight Liberators, of which three were badly damaged and made emergency landings. They were followed by a Boston, whose undercarriage collapsed after it made an emergency landing—this was just one of 199 hectic landings made during the day.

There were nineteen landings during the night of the 14th. and among them was a Mosquito from 107 Squadron, whose pilot had called for emergency homing at 1.45 a.m. Eleven minutes later, he stated that he would have to make a belly-landing with the bombs still on board. At 2.05 a.m., the aircraft touched down on the runway and broke in half, but the crew were not injured and in a short while the aircraft removal party, the MO, and the engineering officer had everything under control.

Some days were quieter than others. For example, on the 17th there were no operational visitors or crashes and just fifty-nine landings were made during the day. They included one Halifax, two Marauders, and two Mosquitos.

Two days later, 143 (Canadian) Airfield Wing, made up of 438, 439, and 440 Squadrons, was diverted from an operation to land at Manston, refuel, and have its pilots briefed for a special raid on a village near Ghent, where there was a report of fifty German tanks based there. One of the Typhoons from 438 Squadron experienced some technical problems after taking off and, after jettisoning its bombs into the sea, the pilot returned to Manston and made a successful belly-landing on the grass.

There were fifty landings on the 16th, but the subsequent days were bleak and there were no operational landings because it was claimed the bad weather was affecting flying. On the 19th, the quiet spell was broken and there were 136 landings during the day, among them was a Marauder that landed on one engine. There were several other emergency landings, including eight Mustangs, two Lightnings, a Fortress, and a Liberator. That night, a Lancaster with a damaged tailplane and port wing made an emergency landing. It had collided with another Lancaster over the target area and one member of the crew was injured.

Two Marauders arrived during the day on the 20th, both being badly shot up, with the rear gunner in the first badly wounded. The second Marauder that had also been badly shot up made an emergency landing with its starboard engine knocked out and a burst tyre. Two of the crew had baled out over the Channel, 5 miles off the coast of Dieppe, believing that the pilot was going to ditch the aircraft, but he had changed his mind and headed for Manston.

On the same day, sick quarters was honoured to receive a visit from Air Marshal Sir Harold Edward Whitingham, the Director General of RAF Medical Services. The son of a Naval officer, Engineer-Rear Admiral Edward Whitingham had studied medicine at Glasgow University. On the outbreak of the First World War, he had joined the Army Medical Corps before re-mustering to the RFC in 1918. At the beginning of the Second World War, the air marshal had been knighted, and in June 1943 he had returned to Glasgow University to accept an honorary degree—he had been told specifically to wear dark clothes for the occasion. In the end, because of the wartime conditions, he had worn his RAF uniform with his academic regalia over the top and nobody complained.

Air Marshal Whitingham was accompanied on his visit to Manston by General Crow of the American Air Force. When they arrived, there was a bit of a mix up after they had requested to see the 'dispersal' sick quarters at Westgate. This was the overspill for the sick quarters at Manston, and it was capable of accommodating thirty-five beds; however, when the air marshal and general turned up, there were no patients. The WAAF Senior NCO explained that she had had to evacuate all the patients to the sick quarters at Manston because there was a German mine floating in the bay that was liable to explode at any time. The air marshal accepted tea and biscuits to the amusement of the American general, who was said to have been impressed by the exhibition of 'British phlegm'.

The sick quarters at Manston was under the control of Squadron Leader Ian Bruce Kerr McGregor. During the years immediately before the war, it had been a station hospital, but, because of budget cuts and lack of use, the building had been used for other purposes. As a result, by 1944 it was hardly fit for purpose and so a new casualty reception centre was built with a sixty-bed hospital, giving Manston sick quarters the ability cope with the expected number of casualties. As previously mentioned, there was also an overspill site in Westgate with another thirty-five beds. The Americans had sent in a crash team of a medical officer and four orderlies, together with a field ambulance and field equipment, to deal with its wounded aircrew.

Sixteen Beaufighter Mk Xs from 143 Squadron were posted to Manston from North Coates on the 23rd to take part in anti-shipping strikes. A number of B-17 Flying Fortresses and B-24 Liberators arrived on the same day, with additional crews to fly out those aircraft that had landed with either technical problems or battle damage. The aircraft had been repaired on site by the American contingent, with the help of the RAF Servicing Wing from 86 MU. There were only sixty-nine landings on the 23rd, a low number that was blamed on bad weather conditions.

The following night, 605 Squadron flew a number of patrols, but lost one of its crew when Flt Lt Parker failed to return—there was no information immediately available about their fate. A couple of Typhoons from 137 Squadron were scrambled to deal with a few hostiles operating in the area of the South Foreland, but they failed to make contact.

During the day of the 25th, there was a lot of activity and landings by 108 Spitfires and seventy Typhoons, plus a number of emergency landings and various prangs. The first was a Marauder from the 454th Bomb Squadron, 323rd Bomb Group, which had been badly damaged by flak and one of its crew had been killed. Another member of its crew had baled out over Ramsgate and was lucky to escape with just a badly bruised leg. The pilot, Captain Robert L. Kelly, made a belly-landing on the east–west grass runway; although a number of the crew were injured, there were fortunately no fatalities.

There was then another Marauder in trouble with one engine out, followed by a Liberator with two engines not working. Later, according to the ORB, a B-17, flying on just two engines with a lot of flak damage to its mainplane, landed in 'Great Syle'. The final prang of the day was that made by a P-38 Lightning, which also had an engine out.

The 28th was a hectic day, with a Marauder of the 394th Bombardment Squadron making a belly-landing and bursting into flames, with one of the crew being badly burnt and another two injured. Following this, a Mosquito of 605 Squadron crashed on landing, while a Typhoon crashed on take-off., when Mosquitos from 605 Squadron were returning from a sortie over Heligoland, claiming to have destroyed three Ju 52s.

The total number of landings in May was 5,148, an increase of 1,977 over April's total. The total number from operations was 4,198, which it was noted was quite good considering that there were ten days of the month when weather conditions had restricted the number of operations. A total number of sixty-two airfield wings, equivalent to approximately 150 squadrons, had arrived at Manston for their crews to be briefed and aircraft refuelled.

The numbers and types of aircraft that landed at Manston, either through battle damage or technical failure, in May 1944 were 2 Lancasters; three Mustangs, fifteen B-17 Flying Fortresses; one Tempest; eight B-24 Liberators; six Typhoons; nineteen B-26 Marauders; ten P-47 Thunderbolts; four Bostons; nine Spitfires; seven Mosquitos; one Swordfish; two Lightnings; and one Martinet. Of these, two B-17 Flying Fortresses, one Mosquito, one Mustang, and one Typhoon landed outside the camp area, but were dealt with by Manston.

The following British and American aircraft were repaired at Manston during May 1944: three Halifaxes; ten B-17 Flying Fortresses; seven Mosquitos; two Liberators; five Typhoons; seven Marauders; five Spitfires; ten P-47 Thunderbolts; and one Tiger Moth. The Tiger Moth is something of a mystery, but presumably it belonged to Training Command or it was being used as the station 'hack'. It certainly seemed a little out of place among the other operational types.

Operations at the beginning of June were badly affected by the weather, with only fifty-three operational landings taking place on the 1st, fifteen operational landings and ninety-four non-operational on the 2nd, and 132 on the 3rd. There were 110 landings on the 4th, twenty-four of which were operational landings and that night there was a show on the camp: H. M. Tennent's *George & Margaret*—a show that had allegedly caused a lot of amusement in the local towns.

On the eve of D-Day, 605 Squadron flew eighteen Intruder sorties, but unfortunately one Mosquito failed to return and Flt Lts Wilton-Brown and Brewis were listed as missing. One of the Wellingtons of 415 Squadron that was equipped with the 'Leigh Light' diverted to Manston after an uneventful Channel patrol and a signals failure. The unit was still operating Albacores in tandem with the Wellington, with the lights of the latter catching U-boats out in the dark, so that they could be seen and attacked by the Albacores. The unit would soon re-equip totally with Wellington Mk XIII.

Tuesday 6 June was described in the ORB as a 'momentous day' and the Station Commander announced over the tannoy that 'the second front has started; from now on [they] may expect greatly intensified activity on all sides'. The total number of landings on this day was only 136, but the weather conditions were again blamed for preventing many aircraft from taking off.

Among the action was 605 Squadron; its Mosquitos flew ten Intruder sorties over enemy held French airfields and claimed to have destroyed a Ju 88, two unidentified aircraft, and attacked and damaged a train. It was noted in the ORB that 'in spite of very poor weather conditions prevailing (one is up for most of the night to assist aircraft in locating the airfield) a constant stream of aircraft taking off and landing'.

On the day after D-Day, the main activity was carried out by the Swordfish of 819 (FAA) Squadron, who flew a total of thirty-four successful sorties for Coastal Command, laying smoke across a fleet of ships sailing across the Channel. That night, there were two emergency landings by Lancasters, the first one from 582 Squadron, which had its elevators jammed and the mid-upper gunner injured, and he second one from 514 Squadron, which had a large hole in its starboard wing where it had been hit by flak and the rear turret had been smashed, with the gunner badly injured. No. 605 Squadron flew eighteen sorties, but one crew, Flt Lt Gathercole's and Warrant Officer Wetton's, failed to return.

Two sections of 137 Squadron were scrambled on the 9th to intercept a number of enemy aircraft that were lurking in the estuary of the River Stour, but the weather prevented any interceptions being made. Eleven sorties were flown by Coastal Command aircraft, but all those arranged for 605 Squadron had to be cancelled due to the weather.

On Saturday 10 June, another pilot was lost. This pilot had was closely connected with RAF Manston and had even acted as Commanding Officer in the absence of Wg Cdr Raphael. The pilot was Wg Cdr John Michael Bryan, who had been promoted in May and, after handing over command of 198 Squadron to Squadron Leader Niblett, had taken command of 136 Airfield Wing. The wing commander's Typhoon was shot down 2 miles south-east of Falaise and he was killed instantly.

Wg Cdr Bryan, one of four sons of the vicar of Milton Earnest, Bedfordshire, had joined the RAF in 1940 and had trained and learned to fly in Canada, being commissioned as a pilot officer on 22 June 1941. Within a short while, he had been posted to 137 Squadron, which was equipped with the Westland Whirlwind that excelled in its ground-attack role. On 23 April 1943, he had been awarded the DFC and in July a Bar was added to it; however, he soon left 137 Squadron to become the Commanding Officer of 198 Squadron and was strongly associated with RAF Manston.

On two separate occasions, 23 and 27 September 1943, Bryan had returned from a sortie with his aircraft badly damaged and certainly on the latter date he had been very lucky to get back. Two other Typhoons had been shot down and Bryan's aircraft was hit in the ammunition tray, leaving a large hole in the mainplane, leaving the aircraft virtually uncontrollable. Only with the use of both arms and one leg was he able to maintain control and fly straight enough to return to Manston.

On the day he was killed, Wg Cdr Bryan was flying as part of 140 Wing and leading 164 Squadron, which was based at Thorney Island and equipped with the Typhoon. What is truly remarkable is that, at the time of his death and having achieved so much, he was only twenty-two years old. Wg Cdr Bryan is remembered on a plaque near the French village of Saint Nicholas, near Vignats, south-east of Falaise. In addition to this, he is mentioned on the Typhoon Memorial near Noyers-Bocage.

Many of the sorties flown during the period immediately after D-Day were anti-shipping sorties, protecting the vast fleet of ships from retaliatory attacks by the Luftwaffe. On the 11th, Typhoons of 137 Squadron destroyed three E-boats and damaged four more. The Beaufighters of 143 Squadron Manston were also active on this day, flying anti-shipping sorties; one aircraft had to make a belly-landing on its return because its starboard wheel would not go down and its port wheel would not go up. It was so badly damaged that, when it touched down on the grass runway, both engines fell out. There were 119 landings on the 11th.

Bombing up aircraft was a dangerous business and although accidents on the ground were not a regular occurrence, there was usually a loss of life when they did. On 13 June, bombs were being loaded onto a Mosquito (NS892) from 605 Squadron, when flames were seen coming from beneath the aircraft; moments later, whatever it was caused the bombs to explode. Four airmen involved in the procedure were killed instantly and one of them was twenty-one-year-old LAC Roy Eastwood Townley from Salford. He was among those that were buried locally in Margate's St John's cemetery.

No less than forty-nine Mustangs arrived during the early morning of the 13th, with one of them having to make a belly-landing due to technical trouble. The Mustangs were followed by a Liberator that still had eight bombs on board, one of which fell off when the aircraft made an emergency landing. The bomb exploded, but nobody was injured and an operation to fill in the crater on the runway began immediately. The Swordfish of 848 (FAA) Squadron moved out of Manston on the 13th, flying to its new base at Thorney Island.

The following day, there was a similar incident to the one with the Liberator. A Beaufighter from 143 Squadron landed at the station with bombs still on board, although the crew were unaware of it. On landing, one of the bombs dropped off. A second bomb then exploded and the aircraft burst into flames, but somehow the pilot and navigator managed to escape with minor burns. Debris was scattered all over the runway and the explosion left another crater to be filled in, some 4 feet square.

There was some excitement on 15 June, when Margate police got in touch with Manston about a number of reports that enemy parachute troops had been seen in the area. The report was passed on to the local Home Guard and the RAF Regiment organised picquets to guard the airfield, but Margate Police soon rang to say that the report was a false alarm and had been discredited.

The first V-1 flying bomb was launched by the Germans on 13 June. Two days later, personnel at Manston had their first experience of the weapon when one of them flew across Thanet—it was so close to the airfield that it was engaged by the Manston guns. No. 605 Squadron's Mosquitos were also engaged in attacking the V-1s during the night of 15–16 June, claiming to have destroyed three of them, as well as a Me 210 and Fw 190. One of them was claimed by Flt Lt J. G. Musgrave and Sgt F. W. Samwell, and it was noted as the first V-1 to destroyed by an RAF pilot. It is interesting to note that, in the Manston ORB, the V-1 Flying Bombs were initially referred to as 'PAs' (pilotless aircraft). Over the next few months, squadrons based at Manston were to play a vital role in fighting the V-1 menace.

During the afternoon of the 17th, another V-1 flew over Manston and was engaged by the guns, but without any result. Mosquitos of 605 Squadron engaged a number of them and claimed to have destroyed two out of a total of six V-1s that were in the general area. Another B-17 (43-37757, *Panhandle*), under the command of Lieutenant McFarlene from the 561st Bomb Squadron, crash-landed at Manston; fortunately, all the crew survived. On the same day, personnel at Manston participated in another 'Salute the Soldier' parade in both Birchington and Margate and everything went according to plan.

The following day, there were more reports of 'pilotless aircraft' in the area and one was reported to have come down 12 miles south of Manston, near Dover. Seven squadrons of 'Bombphoons' landed at Manston on the 20th after carrying out attacks on V-1 sites in the area of Pas-de-Calais. After refuelling and further briefings, they carried out more attacks during the afternoon. That night, two Mosquitos from Hunsdon landed short of fuel after carrying out glider bomb patrols, during which they claimed to have destroyed one of them.

Bomber Command began to make good use of the facility at Manston and, during the night of the 22nd, a Halifax from 78 Squadron made an emergency landing, with one of its starboard engines unserviceable. It had also suffered a lot of damage to its fins and rudder as a result of a collision with another bomber. Only in recent years has it been appreciated how many bombers suffered damage by collisions with other bombers and also hit by bombs from other aircraft flying above them. Lancasters normally operated at higher flight levels than the Halifax (with the exception of the Halifax Mk III with Bristol Hercules engines) and they were more vulnerable.

During the early evening of the 24th, there was a remarkable feat of airmanship by a pilot who landed a Halifax on a single engine. There is no mention of what squadron or type of Halifax this was, but it was almost certainly a Halifax Mk III. The Halifax was followed by a Lancaster that landed with both port engines unserviceable. No. 605 Squadron was heavily involved with anti-diver patrols and one of its Mosquitos made an emergency landing on this night after firing at and destroying a V-1. The pilot had closed in to a range of just 200 yards and the subsequent blast had badly damaged the aircraft's engines, with the debris puncturing a radiator, and so it landed on one engine.

No. 605 Squadron suffered a tragic accident on 26 June, when Mosquito NS880, which was returning to Manston at about 1,000 feet, broke up in the air and crashed close to Margate railway station. The pilot, Flt Lt John Reid, DFM, and navigator, Plt Off. Philips, stood little chance of abandoning the aircraft and were killed instantly.

It was another bad day for the USAAF on 26 June, when five P-47 Thunderbolts were lost and four of the five pilots killed in action. The aircraft were all part of the 61st Fighter Squadron, 56th Fighter Group, and had taken off from Manston at 6.12 p.m. First Lieutenant Ralph A. Johnson was the only one to survive and, after baling out, he was picked up by an ASR launch.

Anti-diver patrols had to be curtailed for a couple of days because of bad weather, but the emergency landings continued with two Liberators on 27 June, both badly damaged by flak. That was followed by three Avengers and two Halifaxes that were suffering from a combination of flak damage and technical trouble. During the evening, a 'diver' was reported flying directly over the airfield; although the guns opened fire, they failed to bring it down.

At 2.10 a.m. on the 28th, an unidentified aircraft was seen approaching the airfield from the direction of Ramsgate. The aircraft was on fire and it was not until it landed that it was recognised as a Mosquito. As soon as it landed on the emergency runway, it burst into a mass of flames, but remarkably both the pilot and navigator managed to escape from what was described in the ORB as a 'holocaust'. Having suffered serious burns, both the pilot and navigator were taken to the sick quarters, but everyone was amazed how they had got out alive in the first place.

On the last day of June 1944, five Avengers, eight P-47 Thunderbolts, thirteen Mustangs, two P-38 Lightnings, and five Mosquitos landed at Manston. No. 137 Squadron carried out a number of anti-diver patrols and claimed to have destroyed two.

The total number of landings in June was 4,004, 2,350 of which were classed as operational landings, with 817 of them made at night. It was noted that the weather during the moonlight period had been exceptionally unfavourable and there had been very few days where there had not been low cloud. No. 155 Wing (General Reconnaissance, FAA) was praised for its work and had been kept very busy on most nights.

A new type of operation had been introduced in June, with 137 Squadron operating against 'pilotless aircraft' by day and 405 Squadron by night. No. 605 Squadron had been credited with thirty-six V-1s destroyed, while 137 Squadron had claimed eleven.

It is strange that there is no mention in the ORB of the FIDO system that was scheduled to be operational from the end of May. There are some references to at least one aircraft, a Short Stirling, landing using FIDO in May, but there is no mention of that type of aircraft landing in the records. It was well-documented that there had been a delay in fitting the Haigill burners on the north side of the runway, but, even after that work had been completed, there is still no mention of FIDO in action.

The number of aircraft that landed with battle damage or technical problems in June 1944 consisted of seven Lancasters; one Wellington; four B-17 Flying Fortresses; four B-26 Marauders; eight B-24 Liberators; seven Mosquitos; four Halifaxes; two P-38 Lightnings; thirteen Beaufighters; six P-47 Thunderbolts; four Typhoons; one Miles Magister; and five Mustangs. A Mosquito, a Mustang, and a Thunderbolt crashed outside the boundaries of the airfield, but were dealt with by the station servicing and recovery section.

The number and types of aircraft that were repaired or disposed of at RAF Manston in June 1944:

British		American	
Lancaster	6	B-17 Flying Fortress	5
Halifax	3	B-24 Liberator	5
Boston	1	B-26 Marauder	4
Mosquito	3	Douglas Boston	2
Hurricane	1	P-38 Lightning	1
Typhoon	3	P-47 Thunderbolt	4
Spitfire	2		

20

A Safe Haven

It was noted that the commencement of July was one of the slackest that Manston had experienced for some considerable time and there were just forty-five non-operational landings. These included two Thunderbolts with engine trouble and three Avengers that arrived from Hawkinge. That night, 137 Squadron was operational on anti-diver patrols along with the Mosquitos of 605 Squadron; the only other activity was by an Albacore of 415 Squadron, which went off on a shipping-reconnaissance sortie.

Over the next few days, it was much of the same, with a flak-damaged Lancaster making an emergency landing on the night of the 4th with all the crew being safe. The next day, two Bostons made emergency landings, one noted as being from the USAAF's 640th Bomb Squadron and the other from the 641st Squadron. However, it is most likely that these aircraft were Douglas A-20 Invaders and part of the 409th Bomb Group, which was based at RAF Little Walden, Essex. No. 137 Squadron was again active on anti-diver patrols and claimed to have destroyed three of them.

There was increased activity on 6 July, with a B-17 making a landing downwind on a single engine, one of ninety landings during the day. That night was one of the most successful to date in the destruction of the V-1 flying bombs, with 605 Squadron claiming six of them. Wg Cdr Raphael, the Station Commander, joined 137 Squadron for the night and claimed to have destroyed one of them, while the unit as a whole destroyed eleven.

A Lancaster from 505 Squadron made an emergency landing during the night of 7 July after being hit by two bombs from another Lancaster that was flying above it during a raid on Caen. The aircraft had been badly damaged behind the port engine nacelle and aileron and, with the pilot struggling to maintain control, it had then been struck by flak that carried away its aerials; this meant that its R/T, W/T, and hydraulics were all out of action. The pilot skilfully landed the Lancaster on a single wheel and, despite all the battle damage, none of the crew were injured.

Apart from a number of other emergency landings and anti-diver patrols, very little out of the ordinary happened over the next few days, but, at 2.40 a.m. on the 21st, there was an unusual occurrence; two enemy aircraft (Bf 109s) appeared over Manston and, after flashing their downward recognition lights, approached the airfield and landed. The first one, *White 16*, flown by *Leutnant* Horst Prenzel, made a normal wheels-down landing, but the second one, *Yellow 8*, with *Feldwebel* Manfred Gromill, overshot the approach and made a belly-landing that badly damaged the aircraft.

Both pilots, who were described as cheerful and happy to give themselves up, had taken off from Saint Dizzier at 12.45 a.m. on a wild *sau* mission over the invasion area. It was an operation by the Luftwaffe's single-engined night fighters that searched the bomber stream for its victims without the use of radar. It was reported that, not for the first time, both pilots had mistaken Manston for their own home station in France, suggesting that there must have been some similarity between the two airfields.

The aircraft that landed intact (*Werk Nummer* 412951, *White 16*) was flown out of Manston on 26 July by Wg Cdr Roly Falk to the Royal Aircraft Establishment at Farnborough. It was later allocated to 1426 Flight, who flew and evaluated enemy aircraft. It was given the RAF serial number TP814 and then delivered to the Air Fighting Development Unit at Wittering, where it was used in tactical trials against various marks of Spitfire and Mustang. The Bf 109 was written-off in an accident at Wittering on 23 November 1944, after it crashed while taking off, but the pilot escaped unharmed.

On 21 July, history was made at Manston when the RAF's first ever jet-fighter unit, 616 (South Yorkshire) Squadron, arrived, equipped with the Gloster Meteor, powered by Rolls-Royce Welland jet engines. The event went almost unnoticed in the ORB. The author simply stated: 'During the afternoon 616 Squadron arrive, and will now be based here'. There is no mention of the Meteor being the first jet into operational service or anything else about this historic occasion.

No. 616 Squadron had been formed in Doncaster in 1938 and for most of the war had flown various marks of Spitfire, being based at a number of Fighter Command's airfields including Hawkinge and Kenley. In May 1944, the Squadron had moved from Fairwood Common to Culmhead in Somerset. On 12 July, the Squadron exchanged its Spitfire Mk XIIs for the Meteor Mk I. At Culmhead, under the command of Wg Cdr A. McDowall, DFM, 616 Squadron acquainted themselves with the jet and tested the Meteor prior to moving and becoming operational at Manston. For a short while after arriving at Manston, 616 Squadron continued to operate a small number of Spitfire Mk VIIs.

Wg Cdr Andrew McDowall began his service in the RAF as an aircraft hand with 602 (City of Glasgow) Squadron before re-mustering as airman untrained pilot. McDowell was called to full-time service with the RAF in August 1939, and after hostilities had begun, he had destroyed a He 111, and a year later went on to be credited with a number of other enemy aircraft. He had been commissioned in November 1940 and awarded the DFM, which was *Gazetted* on 17 December. After becoming a Flight Commander on 245 Squadron, McDowall was posted to a number of other units and his first command had been 232 Squadron, when it reformed with Spitfires at Atcham. Following this, he had been given a desk job in 13 Group before being appointed as the CO of 616 Squadron. It is fitting that the CO of the RAF's first jet unit had come through the ranks.

The pilot of a Spitfire from 1401 Meteorological Flight found a new role on the 25th, when he guided a Lancaster from 49 Squadron to the airfield after it had suffered significant battle damage. The Lancaster was missing both propellers from its starboard engines, but the pilot managed to land safely and none of the crew were injured. The targets on the 25th for Bomber Command were an airfield and signals depot at Saint-Cyr and flying bomb sites at Watten. Of the 175 Lancasters that were deployed, only a single aircraft was lost.

A Tempest of 3 Squadron was diverted to Manston on 27 July because of bad weather and R/T failure. It was the bad weather that was responsible for an incident on the same day, when two Typhoons of 137 Squadron collided. The aircraft were climbing in cloud on an anti-diver patrol when the collision occurred and both Fg Off. Johnson and FS Hack were killed immediately.

A number of sources state that the first operational anti-diver patrol flown by the Meteors of 616 Squadron was on the 27th, but there is no mention of any activity by the unit on that day in the ORB. Despite this, however, there is ample evidence to prove that 616 Squadron's first operational sorties did take place on that day and the Commanding Officer Wg Cdr McDowall was among those who were airborne. He flew in Meteor EE222, which he had adopted as his personal aircraft and had decorated it with the pennant of his rank and Canadian flag. Fg Off. Bill McKenzie also flew his first sortie on this day.

The unit is first noted in the ORB as being operational on the 28th, taking part in 'diver' (anti-diver) patrols with the Typhoons of 137 Squadron, but without any results. The following day, 616 Squadron was active again with 137 Squadron on what was described as 'abortive diver patrols'.

The total number of landings for July was 4,156, of which 2,285 were operational landings—861 of them were made at night. No. 605 Squadron was credited with destroying three enemy aircraft and twenty-five flying bombs, while 137 was credited with seventeen.

A note by Wg Cdr Raphael in the ORB stated: 'sick quarters have carried out their usual good work in connection with the crashed aircraft, tending 9 American aircrew and 4 RAF'. The total number of landings at Manston due to battle damage, technical failures, or fuel shortage was 282.

A Safe Haven

The aircraft that crashed or landed at Manston with battle damage and or technical problems in July 1944 included one Lancaster; five Beaufighters; six B-17 Flying Fortresses; one Gloster Meteor; nine B-24 Liberators; seven Typhoons; one Wellington; one P-51 Mustang; four B-26 Marauders; one Spitfire; two Bostons; two P-47 Thunderbolts; and one Mosquito. The following table shows American and Allied aircraft that were repaired at RAF Manston during July 1944:

Allied		American	
Lancaster	4	B-17 Flying Fortress	6
Halifax	2	B-24 Liberator	9
Mosquito	3	B-26 Marauder	7
Typhoon	3	P-51 Mustang	1
Spitfire	1	P-47 Thunderbolt	4

On the first day of August, forty-seven Spitfires of 132 Airfield Wing, Tangmere, landed after providing cover over the target for a force of Lancasters that were attacking a position near St Omer. The target, described as 'special', was a number of flying bomb and storage sites; out of the force of 385 Lancasters, only seventy-nine bombed the aiming point because of bad weather.

A section of Spitfires from 616 Squadron carried out reconnaissance sorties over northern France, but the weather interfered with the sorties and one aircraft landed back at Exeter. No. 819 (FAA) Squadron moved out of Manston on the 1st and flew out to its new base at Inskip. In spite of the bad weather, there were 122 landings on the first day of the month.

On 3 August, two Liberators landed at Manston, the first one being described as being in 'dire straits', with its port engine knocked out and five of its crew having baled out over France. As soon as it touched down, the aircraft burst into flames and, despite the best efforts of the crash crew, which was on the scene straight away, two of the crew were trapped and killed. Another two managed to escape the carnage and it took rescue crews over an hour to completely get the situation under control. The second Liberator had both starboard engines unserviceable, but its pilot managed to make a safe landing.

21

Squadron Leader Joseph Berry: The V-1 Champion

On 2 August, 501 Squadron had arrived at Manston with its Hawker Tempest, under the command of Squadron Leader Gary Barnett. Nine days later, he handed over command to Squadron Leader Joe Berry (118435). Like Wg Cdr McDowell, the CO of 616 Squadron, he had risen through the ranks and was a very experienced pilot. Joe Berry had enlisted in the RAFVR during August 1940 and trained as a fighter pilot, after which he had been posted to 256 Squadron at Squires Gate, Blackpool.

No. 256 Squadron was then equipped with the Boulton & Paul Defiant and was assigned in the night-fighting role for the defence of north-west of England, and there were plenty of accidents. On the night of 4 November 1941, Sgt Berry was taking part in a practise 'fighter night', when the oil pressure in the Merlin engine of his Defiant, T4053, suddenly dropped.

The then-Sgt Joe Berry ordered his air gunner, FS Williams, to bale out and, although it was quite difficult to abandon a Defiant, he soon followed him, baling out and landing safely. FS E. Vivian Williams was not so lucky and, probably due to a gust of strong wind, he was blown offshore and landed in the sea. Williams shouted for help and his calls were heard above the noise of the wind and sea. The Fleetwood lifeboat was launched, but bad visibility prevented him from being rescued and he drowned, with his body being washed ashore the next day. This would not be the first time that Joe Barry would have to abandon an aircraft.

Joe had been commissioned in March 1942 and soon afterwards he married Joyce, who he had first met while working for the Inland Revenue in Nottingham. They did not have the chance to spend a lot of time together and soon after moving to Woodvale with 256 Squadron, with his friend and colleague, Plt Off. Bryan Wild, Joe was sent to Filton at Bristol to collect a brand new Beaufighter. After a number of fuel consumption test, Joe and Bryan, together with their navigators flew to Gibraltar and then on to Serif in the Atlas Mountains. There Joe and Bryan were put into the '*Pilot Pool*' along with a number of others and after three weeks he posted to 153 Squadron that was equipped with the Beaufighter.

Squadron Leader Joseph Berry: The V-1 Champion

Joe was later posted to 255 Squadron, equipped with the Beaufighter V1f, and it was while he was with that unit that he opened his account on 9 September 1943, when he destroyed a Me 210 south of Capri and another one near Largo-Paolo the following day. On 3 October, Joe took part in what was later known as the 'Great E-boat Raid' against a German invasion force attacking the Greek island of Cos. Sixty Beaufighters set out on the operation, but, because of heavy ground fire and strong headwinds on the return flight, only twenty-five returned safely. On 24 October, Joe claimed his third enemy aircraft, a Ju 88, south of the Volturno River.

In February 1944, Fg Off. Joe Berry returned to England. In March, he was awarded the DFC for his service in the Middle East. Wg Cdr Hartley, who commanded the fighter interception unit at Tangmere, knew Joe and asked for him to be posted to the unit; he arrived there on 25 June, having recently been promoted to the rank of flight lieutenant. It has been claimed that he had been chosen because of his unassuming demeanour and there was a lot of competition to get into the FIU as it had a reputation as being a 'crack unit'. One of the many different types of aircraft that he flew at the FIU was the Westland Welkin, a twin-engined fighter designed to fly at very high altitudes and built with a pressurised cockpit. Only seventy-seven Welkins were ever built, and the FIU had two of them, which mainly operated from Wittering.

The FIU was still part of the ADGB and in June it received a request to convert some of its best blind-flying pilots to the Hawker Tempest; Flt Cdr Sqn Ldr Teddy Daniel and Flt Lt Joe Berry were the first two that were chosen. On 25 June, they were sent to Newchurch, where they became the nucleus of a small detachment and set up camp away from the resident day unit, subsequently suffering a lot of joking and jibes in the officers' mess.

The Tempest was based on the design of the Typhoon and, with its Napier Sabre engine, was capable of speeds well in excess of 400 mph. The early versions had been fitted with 20-mm long-barrel Hispano cannon, but later production models had short barrelled cannon that were flush with the leading edge of the wing.

Joe Berry destroyed his first V-1 on the night of 28 June, in fact he was credited with two of them on that night while flying on patrol with Squadron Leader Daniels; over the subsequent nights, he had further success, destroying three on 30 June, another two on 5 July, and four on the night of the 6th. On 23 July, he set a new record by destroying seven of them in a single night. However, attacking the V-1s was a dangerous business and on the 27th his aircraft was damaged while chasing another one near the RAF airfield at West Malling.

According to the Manston ORB, there was no activity by 501 Squadron on the night of 12–13 August. The ORB stated: 'Two Meteors of 616 Squadron fly anti-diver patrols, but see nothing'. That night, however, Squadron Leader Berry filed a 'diver report', describing how he destroyed two of them. We acknowledge and appreciate the help of Graham Berry for providing such detailed information.

Left: Squadron Leader Joe Berry, the Commanding Officer of 501 Squadron, who was credited with the highest number of V-1s destroyed—a tally of sixty.

Below: Hawker Tempest EJ538 with a number of pilots from 501 Squadron. This photograph was taken shortly after the unit had moved from Manston to Bradwell Bay. Squadron Leader A. Parker Rees, DFC, who had taken over from Squadron Leader Berry, is in the centre with his hands in his pockets.

Consolidated Diver Report No. 8. 501 Squadron RAF Manston.
Time on Patrol (1) 0005–0215

1. I was on patrol under Watling control when I saw a Diver coming from the direction of Rye at 2,000 feet at 400 mph on a course of 330 degrees. I immediately closed in and, from 150 yards astern, I opened fire. I saw strikes on the motor and the Diver crashed north of Sandhurst at 0123.

Time on Patrol 0410–0630

2. On my second patrol under Watling control the weather at base was reported as raw. I flew over, headed to divert to Ford and on the way there was trade reported coming in from Dungeness. I saw one Diver and turned in behind it (position uncertain). Height was 1,600 feet, course 340 degrees speed 190 mph. I fired two short bursts from dead astern range 150–300 yards. No result. Third burst from 250 yards and the Diver exploded in the air. Position was approximately 6 miles southwest of West Malling. Watling have a fix (claim is in daylight).

Consolidated Diver Report No. 10 [19 August]. Up 0615 Down 0727.
I was scrambled for Divers reported coming in and was controlled by Biggin Hill, two were reported coning in over Dungeness, I was 6 miles north of Rye when I saw a Diver coming in just over cloud at 2,400 feet on course 330, at 300 mph. I attacked it from astern firing from 200 yards' range with two burst, the Diver exploded in the air at 0635 hours and crashed at West Malling.

Consolidated Diver Report No. 11 [31 August]. Patrol from Manston 0540 until 0645.
I was patrolling under Watling control over Ashford and to the west when I saw a Diver in the Sandwich area at 3,000 feet on a course of 290–300 degrees at 250 mph at 0645. I closed in to 300 yards dead astern and fired a short burst which knocked pieces off the propulsion unit, I fired again from 150 yards and saw more strikes, the Diver exploded on the ground in the Faversham area at 0550 hours.

No. 501 Squadron's time at Manston was relatively short. On 22 September, the Squadron was moved to Bradwell Bay, Essex, to counter the threat from air-launched V-1s from He 111s that were operating from Dutch airfields. Squadron Leader Berry and 501 Squadron continued to search for and destroy V-1s and, by 4 October, the day that he was killed, his tally had risen to sixty V-1s destroyed. To put that into context, the next highest tally was by Flt Lt Mellerish of 96 Squadron, who was credited with forty-seven. Wg Cdr Roland Beaumont of 609 Squadron and Group Captain Johnson were both credited with forty-one, while Wg Cdr Crew, CO of 96 Squadron, was credited with thirty-four.

22

The Deadly Menace

On 4 August, 616 claimed their first V-1 flying bombs destroyed, one of them in the conventional manner (with the use of guns) and the other by a more unorthodox method. The first one was claimed by Fg Off. Rogers, who used his aircraft's guns, but Fg Off. Dean guns jammed as he approached his intended target; not to be outdone, he flew alongside the V-1 and tipped it over with his wing tip. Very few aircraft had the speed to keep up the pilot-less bombs, but the Meteor was one of them. During the evening, it was noted that Fg Off. Somes of 137 Squadron also destroyed a V-1 on this night using guns.

At 3.30 a.m. on the 5th, low stratus cloud made flying very difficult and all available lighting was switched on to aid incoming aircraft The cloud base was down to 200 feet and the situation lasted until dawn, but fortunately there were no emergencies during this period.

Emergency landings were made by a Boston, a Lancaster, and a Halifax on the 5th, with the two 'heavies' having carried out attacks on flying bomb sites in the Pas-de-Calais area. The Lancaster from 578 Squadron landed on three engines, while the Halifax had two of its engines badly damaged. The Meteors of 616 Squadron flew uneventful 'diver' patrols, along with the Tempest of 501 Squadron, which destroyed two of them.

Sir Archibald Sinclair, the Under Secretary of State for Air, arrived at Manston on 7 August and although there is no mention of the purpose of his visit, it was almost certainly to discuss the threat posed by the V-1 and what was being done. Over the years, the former First World War airman had visited Manston on a regular basis and was a familiar face in the officers' mess. The weather was still interfering with operations and all night flying was cancelled.

Squadron Leader Piltingsrud and the Typhoons of 137 Squadron departed Manston on 13 August for a forward airfield in France, known as B.6 Coulombs, which had been completed on 16 June with a runway that was 1,520 metres long. The unit was replaced by the Spitfires of 504 (County of Nottingham) Squadron, arriving from Detling under the command of Squadron Leader Banning-Lover.

On 15 August, the Harrowbeer and Culmhead Wings arrived at Manston from their bases near Plymouth and Somerset. The pilots of the ninety-seven Spitfires were briefed and their aircraft refuelled before they took off again to escort a force of 1,004 heavy bombers that were to attack German-held airfields in Holland and Belgium in the build up to the resumption of the night-bombing offensive. After the operation, the Spitfires landed back at Manston and, after the pilots were debriefed, they then returned to their own airfields.

During the evening of 15 August, 616 Squadron lost one of its Meteors after it had taken off from Manston for High Haldon Airfield, near Ashford, where it was due to join the readiness section. The aircraft was serial number EE226, flown by twenty-one-year-old Warrant Officer Donald Gregg, and it crashed at an advanced landing ground, Great Chart, Canterbury, after abandoning an approach to another ALG at High Haldon. It was claimed that the Meteor stalled as it had approached the runway and spun into the ground. The death of Warrant Officer Donald Arthur Gregg from Nottingham was the first fatality in a British jet aircraft. EE226 was part of the production batch of twenty Meteors ordered under contract A/C 1490/41 that were give serial numbers between EE210 and EE229.

The following day, Fg Off. Bill McKenzie, who had flown 616 Squadron's first operational sortie on 27 July, destroyed his first V-1, one of two claimed by the unit that day. He was about to land at Manston when he was informed that 'trade' was coming in and he flew to a position 3 miles south-east of Ashford at 3,000 feet, where he spotted the 'diver' flying on a course of 320 degrees at 1,000 feet. He had to compete with a Mustang that suddenly appeared and fired at the 'diver', but failed to hit it. At a distance of 400 yards, McKenzie opened fire with a four-second burst and it was reported crashing 6 miles south-east of Maidstone at 9.40 a.m. Strikes were seen hitting the 'diver' and its one of its wings fell off, rolling it over and causing it to fall to the ground. McKenzie's claim was supported by the Royal Observer Corps, who witnessed the action.

No. 274 Squadron was posted to Manston on the 17th, having recently exchanged its Spitfire IX for the Hawker Tempest V at West Malling. It was the second squadron in what would be Wg Cdr John Basil Wray's Tempest Wing.

No. 616 Squadron was active again on the 19th, flying anti-diver patrols and added another one to its tally. No. 54 Squadron was scrambled to intercept a number of enemy aircraft that were thought to be in the area, but in the end it proved to be a false alarm and the patrols proved to be uneventful.

Flt Lt Cyril Brooking Thornton, MBE, (117692) of 501 Squadron was killed on 20 August while taking part on an anti-diver patrol in Hawker Tempest

EJ602, which crashed at Vine Farm near Eastry, just outside Sandwich. Born in Croydon and educated at Shakespeare's School, King Edward VI School, in Stratford, he had received his MBE for 'brave conduct', which had been published in *The London Gazette* on 14 March 1944 (page 1223).

Like his colleague and Commanding Officer, Squadron Leader Joe Berry, twenty-six-year-old Flt Lt Thornton had served at the Fighter Interception Unit and he had been credited with destroying nine V-1s. He was married and the address of his widow, Beatrice Louisa Thornton, was given as Edgbaston, Birmingham.

Flt Lt Thornton's bravery and his subsequent award of the MBE was for his actions during an incident in which a Mosquito had crashed while taking off and he had been one of the first on the scene. He found the navigator, who had been thrown clear, and, having comforted him, organised the crash crews and medics, making sure that he was taken care of before turning his attention to the pilot. The main area of the wreckage was surrounded by barbed wire and Thornton had found the pilot lying beside an engine with his clothes on fire. After putting the fire out with his bare hands and removing the pilot's parachute, Flt Lt Thornton then helped to put him in an ambulance. Unfortunately, both airmen died later, but it was recognised that had it not been for his actions, both would have died in the fire and not stood any chance at all of surviving.

Bad weather curtailed any activity at Manston on the 21st, and there were just eleven landings during the day. The conditions did not improve during the night and there were no aircraft movements at all.

Five squadrons of Spitfires from No. 10 Group (approximately sixty aircraft) arrived at Manston on the 24th to be refuelled and the pilots briefed for an operation that was then cancelled. No. 274 Squadron claimed a 'diver' and 504 Squadron had a successful day shooting up a number of vessels, small and large, off the Dutch coast. There were 176 landings during the day. No. 504 Squadron had another successful day on the 26th, when it had left twelve trucks on fire, shot up a barge and a locomotive, and damaged a lot of rolling stock.

On the night of 27 August and getting away from the station's operational role, an audience was entertained by an old and respected friend of Manston, Mr W. H. Tennant, who had produced a stage version of the film *Ten Little Nigger Boys*. The film has been remade several times, but has been given a more politically correct title for this modern age. It was noted in the ORB that 'the artist got well and truly operational on two performances—no hitches, and a very fine undertaking'.

Wg Cdr McDowell, the CO of 616 Squadron, was involved in an incident on the 29th, when his Meteor, EE222 YQ-G, crashed after suffering engine failure and running out of fuel—he chose to adopt the standard forced-landing procedure for the Spitfire, 'wheels up'. The aircraft then careered across a field and a number of hedges at Plucks Gutter, which was close to the public house The Dog and Duck. Although the wing commander was only slightly injured, he would have been upset to have lost the aircraft of his choice, his own personal 'mount', which was a complete write-off. On the same day, 80 Squadron, the third and final element of Wg Cdr Wray's Tempest Wing, arrived from West Malling, having recently re-equipped from the Spitfire IX to the Tempest V.

There was another surprise in store for those serving in flying control at Manston on the 30th, when, at 12.55 p.m., a Fw 190 appeared north of the airfield and flew across the Monkton Marshes before making a wheels-up landing in a field on Monkton Road Farm, Birchington. It may have surprised those who first arrived at the scene to find out that the pilot was not German, but Dutch. Johannes Kumn was a Dutch ferry pilot, who had been tasked with flying the latest mark of fighter, the Fw 190A8, to JG 26 based at Melsbroek Airfield, near Brussels. He had taken off from Wiesbaden at 11.30 a.m. and, due to a combination of bad weather and his intention to defect, he decided to fly to England. The Germans should have known better than to trust a Dutch a pilot with the latest model of the Fw 190 as they had a 'track record' for stealing their aircraft.

The last day of August was a sad one for a number of reasons, but mainly the loss of a number of personnel, the first one being Flt Lt Bishop of 504

Fw 190 A.8 that made a forced-landing at Manston on 30 August 1944, when the Dutch ferry pilot decided to defect to Britain.

Squadron, who was killed when his Spitfire, PL222, crashed close to the beach at Sandwich during an air test. Bishop was on his second tour of duty and the usual Spitfire that he flew was PL379. The ill–fated Spitfire PL222 was a relatively new aircraft that had only been allocated to 504 Squadron in June. Flt Lt Kenneth William Bishop, son of Walter S. Bishop and Violet from Birmingham, was buried in Witton Cemetery in his home town.

It was also a tragic day for 605 Squadron when Fg Off.s Brigden and Harris were reported missing after their Mosquito failed to return from a sortie. Nine shells that had been fired by the German guns in France were reported to have landed in Ramsgate and one exploded in the area of the officers' mess at the radar station in Sandwich. The shock wave and tremors from another shell that landed at Haine was felt on the airfield at Manston.

The American connection with RAF Manston was reaffirmed in August, when Captain Samuel Velebny arrived, posted in as the Commanding Officer of Detachment 'A' of the 16th Mobile and Repair Squadron. It had been formed on 25 October 1943 and activated on 1 November that year as part of the Thirteenth Air Force's Air Technical Service Command. The unit consisted of five officers and 250 enlisted airmen, whose role was clearly set out. They were to home in all American aircraft that were suffering battle damage or technical problems to Manston, to co-ordinate operations for the return of American airmen to their base stations, and to handle emergency repairs and refuelling of all American aircraft.

The feeding, re-clothing, and provisioning of American airmen was another of their responsibilities for those who were forced to remain at Manston because of bad weather or any other reason. The detachment had the responsibility for the repair and refuelling of all American aircraft within a 70-mile radius from Manston, plus the care of and evacuation of all personnel who were wounded or injured that landed there.

The tally for the highest number of V-1s in August went to 501 Squadron, who destroyed twenty-nine, well-ahead of 616 Squadron, which was credited with fourteen and 274 Squadron with thirteen. The Mosquitos of 605 Squadron were credited with seven, but also two enemy aircraft, and 137 Squadron just two V-1s.

The aircraft that crashed or landed at Manston with battle damage or technical problems in August 1944 included four Lancasters; 7 Tempests; three B-24 Liberators; three Typhoons; 3 Halifaxes; one P-38 Lightning; one Boston; four P-47 Thunderbolts; two B-26 Marauders; ten Spitfires; seven Mosquitos; one Proctor; one P-51 Mustang; one Magister; and four Meteors. The American and Allied aircraft that were repaired at RAF Manston during August comprised:

Allied		American	
Lancaster	2	B-17 Flying Fortress	1
Halifax	2	B-24 Liberator	3
Mosquito	4	Boston	3
Meteor	1	B-26 Marauder	3
Tempest	3	P-51 Mustang	1
Typhoon	3	P-38 Lightning	1
Spitfire	1	P-47 Thunderbolt	1

The inclusion of the Percival Proctor and Miles Magister seems a little bit frivolous against all the other types, but presumably man hours and expense had been spent working on them that had to be accounted for.

The busiest day for some time proved to be on 1 September, when four Spitfires squadron from the North Weald and Deanland Wing (Deanland was an advanced landing ground in Essex) arrived prior to escorting a number of B-25 Mitchells on an operation over France. Within twenty-five minutes, the crews had been briefed and the aircraft refilled and sent on their way. A Halifax III from 10 Squadron, based at Melbourne, Yorkshire, made a crash-landing on the airfield, having had two of its engines badly damaged during a sortie over Normandy—none of the crew were injured. The Manston Tempest Wing was active, flying anti-diver patrols, but failed to claim any. There were 285 landings throughout the day.

Nos 3209 and 3210 Service Commando Units arrived for duty at Manston on the 3rd, although there is no mention what their exact duty was. These units had been formed on the advice of Lord Mountbatten and its personnel were trained on similar lines to the Army and Royal Marines. Their role was to accompany front-line troops to any active theatre of war and build or repair airfields or landing grounds constructed by the Army Airfield Construction Units; they were also to repair aircraft on the front line, under fire. Each unit had between two and three officers and 150–170 other ranks that were equipped with Jeeps, motorcycles, and 15-ton trucks. Both 3209 and 3210 had been formed in April 1943 and after serving in Normandy both units would later be posted to the Far East.

On the same day, three squadrons of the West Hampnett Wing landed to refuel and for its pilots to be briefed for an operation involving 100 Halifaxes, which were to bomb Volkel Airfield in Holland, where a number of Luftwaffe units were based with the Ju 88, Arado 234, and Me 262. No. 504 Squadron also took part in this operation and Commanding Officer Wg Cdr Raphael took the opportunity to fly with them. Altogether, a total of 675 aircraft took part in this operation, including 348 Lancasters and 315 Halifaxes and six airfields in Holland were targeted. Only a single Halifax was lost.

23

Remembering Arnhem

There was lots of excitement on 5 September, when thirty-seven gliders, full of airborne forces personnel, towed by Armstrong Whitworth Albemarles from 296 and 297 Squadrons, arrived from Brize Norton, in what was described in the ORB as 'no small strength'. It did not take very long for the paratroopers to disembark and it was noted that this was the first time their distinguished red berets had been seen at Manston in such a large number. More gliders and their tugs continued to arrive at Manston over the next few days.

Most of the men were from the First Allied Airborne Army that been formed on 2 August and they were scheduled to take part in Operation Comet, a risky attack behind enemy lines to capture a number of bridges. The Operation, which was to have taken place on 8 September, was cancelled in favour of Operation Market Garden and that left a large number of soldiers with too much time on their hands. As a result, discipline broke down and a series of short 'bull courses' were introduced to reinstate it, to the satisfaction of both the Army and the RAF.

Squadron Leader Wise was replaced as the station's administrative officer by Squadron Leader J. J. Secter, who was posted in from Dunsfold. A Lancaster from 460 Squadron at Binbrook made an emergency landing on two engines after being hit by flak over Le Havre, wounding the navigator. There were 380 landings during the day, but not a single accident or prang.

No. 263 Squadron arrived at Manston on 6 September from ALG B.3 at St Croix, having been re-equipped with the Typhoon in December 1943. Just five days later, it flew out of Manston to be based at B.51 Lille, where it resumed ground-attack missions for the rest of the war.

Heavy rain restricted activity on the 7th, but twenty-six Spitfires from 222 and 249 Squadron arrived from Normandy because their own airfields in France were flooded—they flew out to Bradwell Bay later on. Captain Harold Balfour, the Under Secretary of State for Air, visited Manston again, just over a month since his last visit.

The battle against the V-1 menace was still going on, with the Luftwaffe changing tactics and adapting the He 111 bombers to carry the weapon instead of having to use launch sites. At 6.44 p.m. on 8 September, the first V-2 rocket landed in Staveley Road, Chiswick, London, and it not only shocked everyone, but drastically changed things again because there was no defence against them and both AA guns and fighters were unable to even challenge them.

On 11 September, there were more emergency landings than Manston had dealt with for some time, the first one being a Boston that had a fire in its starboard wing after being hit by flak. This was followed by a Liberator that landed on two engines, then another Liberator that landed on three engines, but had also been extensively damaged by flak—the rear gunner having been killed and a waist gunner badly wounded. A B-17 landed on three engines and, as it touched down, the propeller of the dead engine fell off onto the runway. A Spitfire from Hawkinge landed after the pilot had found out that the cockpit hood would not open; although it might not have been considered to be as serious a problem as the others, it would have prevented him baling out and would have effectively killed him.

The Detling, West Hampnett, and Deanland Wings arrived again on the 13th for the aircraft to be refuelled and the pilots briefed prior to them escorting a force of Halifaxes that were going to bomb Gelsenkirchen. A Liberator that had landed on just two engines had to be towed off the runway because it did not have enough fuel left to power itself; this was followed by a B-17 that was in the same predicament.

Just after 10 a.m. on Sunday 17 September, the first elements of First Allied Airborne Army left Manston in fifty-six Horsa gliders, towed by Armstrong Whitworth Albemarle tugs, which had arrived from Brize Norton over the previous two days. They were part of a force of 1,438 Dakota C.47s and 916 Airspeed Horsa gliders taking part in the ill-fated Operation Market Garden. Glider pilot S/Sgt Shackleton remembered that, as he was checking that everything had been loaded properly, he was approached by a man with a tripod who asked him where 'the best place to film them taking off' was. It turned out that he was from Pathé News, and the film footage that he took that day can still be seen on the Pathé website.

On the first lift were forty-six Horsa gliders, towed by Albemarle tugs from 296 and 297 Squadrons, based at Brize Norton, and another ten American-built Waco CG-4 gliders, also towed by Albemarles. On board the Horsas were men from the 2nd Battalion, South Staffordshire Regiment, and sections of the Field Ambulance and Royal Army Medical Corps, while on the Wacos were elements of the American Airborne Corps and USAAF Air Support Signals Team. Manston was the only airfield from which the Albemarles operated.

The Manston Tempest Wing, led by Wg Cdr Wray, took off and supported the operation by attacking flak position on the Dutch Islands, over which the force was routed. Shortly after midday, the first of five Dakotas arrived at Manston that had taken part in the Arnhem operation, most of which had been badly damaged by flak and needed to make emergency landings. There were 248 landings during the day. It may have been ironic, but on the same day a service was held in the station church to commemorate the Battle of Britain, followed by a march past with the salute being taken by Wg Cdr Raphael.

All the remaining tugs and gliders that remained at Manston were flown to Holland the following day and were again supported by Wg Cdr Wray and the Tempest Wing. Another forty-two Albemarles of 296 and 297 Squadron towed an equal number of Horsa gliders, which carried elements of the Polish Parachute Brigade and the Royal Artillery along with some of their guns. Albemarles and Horsas that took off independently carried the Officer Commanding 'B' Company of the 2nd Battalion, South Staffordshire Regiment.

Not all the gliders made it to Arnhem and one of those on the second lift, no. 878, was hit by 'friendly' flak off the coast near Middlesbrough; however, with the aid of the pilot of the Albemarle, it managed to continue in flight until the tow rope broke. The glider crashed near a Dutch farm and those on board put up a fierce fight against the odds until they eventually surrendered. It was noted in the ORB that forty-two Albemarles returned to Manston after the operation before flying back to Brize Norton.

Spitfires from 118, 124, and 303 (Polish) Squadrons that had formed the West Hampnett Wing arrived shortly before midday on Tuesday the 19th for the pilots to be briefed and aircraft refuelled prior to escorting gliders and their tugs to the area of the Rhine Delta area of Holland. Only a single Albemarle and Horsa glider departed from Manston, carrying elements of the Royal Artillery, and the largest contingent, twenty Halifaxes and Horsas, took off from Tarrant Rushton. Owing to the bad weather, the escort had to be abandoned and the Spitfires returned to Manston where they remained overnight. Later in the day, two Walrus ASR aircraft landed at Manston carrying four members of glider crews that had been shot down in the sea.

On 20 September, the Tempest of 274 Squadron departed from Manston and flew out to its new base at Coltishall. It had only been at Manston for a month, during which time it had covered the Normandy landings. It was destined to join the 2nd Tactical Air Force to fly sweep over the low countries. No. 80 Squadron also left on the same day for Coltishall and it would soon be posted to Belgium to operate armed-reconnaissance sorties.

There is the first mention of FIDO on 21 September, although that does not mean it had not been used before, but maybe just not mentioned in the ORB. On the 21st, it was burned from 12.30 p.m. to 6.45 p.m. to assist in nineteen out of the twenty landings during the day, in what was described as extremely bad weather. Among the different types of aircraft that landed were three Norseman, eight Spitfires, two Mustangs, and a Stirling that had been badly damaged by flak. There was also an Auster that arrived from France, a strange type to be flying around in during very bad weather.

It was also a memorable occasion on the 21st because, after serving as the Station Commander of RAF Manston for seventeen months, Wg Cdr Raphael, DSO, DFC, handed over command to Wg Cdr A. D. Murray, DFC. Raphael was posted to the RAF Staff College and was destined for higher things, having overseen the station through crucial changes and changing times, especially new 9,000-foot runway.

Wg Cdr Alan Duncan Murray (34168), who took over as the Commanding Officer, had been awarded a Short Service Commission in the RAF during January 1934 and he had learned to fly at No. 3 Flying Training School. His first unit had been 18 Squadron, equipped with the Hawker Audax, but he was chosen to fly floatplanes and was posted to Leuchars for catapult training and then to Calshot to convert to floatplanes. He was then posted to Gosport to undergo deck landing and torpedo training, before moving on to HMS *Malaya*, a *Queen Elizabeth*-class battleship, where he had flown the Swordfish floatplane.

By 1939, the Wg Cdr had been posted to the Aircraft and Armaments Experimental Establishment at Martlesham Heath and moved with the unit to Boscombe Down. Shortly after the outbreak of war, he had undergone a flight refresher course at 6 OTU on the Hurricane and later joined 46 Squadron. In September 1940, he had spent a short amount of time with 501 Squadron before being given his first command of 73 Squadron. He had been awarded the DFC in March 1941 before being posted overseas to serve in Egypt and Iran, where he controlled the fighter sectors. Having returned to the UK in March 1944, he had commanded a servicing unit in the south of England and from there he went on to be posted to Manston.

The day after Wg Cdr Murray arrived, the Tempests of 501 Squadron under the command of Squadron Leader Joe Berry left Manston for Bradwell Bay, Essex. Shortly before he left Manston, the squadron leader reprimanded a senior officer for taking one of his Tempests without his permission. It was a dispute that was to have repercussions and one that possibly cost him his life.

There was still a number of sorties being flown in support of the ground operation at Arnhem and, on 23 September, 504 Squadron flew a considerable number of them to escort the Dakotas and Stirlings that were dropping supplies. Two Stirlings made emergency landings at Manston, the first from 196 Squadron with an engine unserviceable and the second also with the engine unserviceable, but having been badly damaged by flak and with the bomb aimer wounded.

That night, a Halifax from 462 Squadron, based at Driffield, and a Lancaster from 166 Squadron at Kirmington landed at Manston after attacking Neuss in the Ruhr. The Halifax had been hit by flak just after crossing the enemy coast and had flown to the target and back on three engines, while the Lancaster had also landed on just three engines. It had been badly damaged by flak, which had destroyed much of its equipment, but then attacked by a night fighter that killed one of the gunners and wounded the navigator and wireless operator. The crew had got some satisfaction when they saw the night fighter that had caused the damage hit by sustained fire from the air gunners and explode on the port quarter.

Three new units were posted in to Manston on 25 September: No. 118 Squadron, under the command of Squadron Leader P. W. E. Heppell, DFC; No. 124 Squadron, under the command of Squadron Leader G. W. Scott, AFC, who had just taken over from Wg Cdr Thomas Balmforth, DFC; and No. 229 Squadron, whose Commanding Officer was South African Air Force Officer, Major Newton Francis Harrison—all three units were equipped with the Spitfire IX. Within a short while of arriving at Manston, 118 and 124 Squadrons joined 504 Squadron in an operation attacking enemy flak positions at Arnhem. By then, with the exception of small pockets of resistance, the battle had been lost.

In just fifteen minutes on the 29th, 100 aircraft had landed, including twenty-five Spitfires from 41 Squadron, nine Tempests from 80 Squadron, and twenty-three Douglas Bostons from 342 and 88 Squadrons. Flt Lt Wood and Plt Off. Johnson were returning to Manston after operating out of Petershead for sorties against targets in Norway, when they spotted and destroyed a V-1 close to RAF Stradishall. There was a total of 199 landings during the day. The aircraft that crashed or landed at Manston with battle damage or technical problems in September 1944 included ten B-17 Flying Fortresses; two P-38 Lightnings; seven B-24 Liberators; thirteen Tempests; four Halifaxes; six Typhoons; six Dakotas; two Mosquitos; three Bostons; three P-47 Thunderbolts; and six Spitfires. American and Allied aircraft that had been repaired at RAF Manston during September comprised:

Allied		American	
Lancaster	8	B-24 Liberator	5
Stirling	1	B-17 Flying Fortress	6
Halifax	2	B-26 Marauder	2
Dakota	4	Boston	1
Mosquito	10	P-51 Mustang	2
Tempest	6		
Proctor	1		
Spitfire	7		
Magister	1		

The first day of October 1944 was noted in the ORB as being 'unusually quiet'. The weather was classed as fair to fine, and there were seventy-two landings during the day. Just after first light on the 2nd, a section of 229 Squadron were sent off on a shipping-reconnaissance sortie to Dunkirk. Landing back at Manston, Plt Off. K. E. Clark was killed when his aircraft crashed on landing and burst into flames.

On the 4th, the weather was still interfering with operations, but 504 Squadron carried out successful shipping-reconnaissance sorties and four Mosquitos from 605 Squadron flew to France to land at forward bases to carry out day-ranger sorties. However, the weather prevented them from landing and they were forced to return to Manston.

The Spitfires of 1, 131, and 165 Squadrons landed at Manston on the 6th for the pilots to be briefed. The aircraft were refuelled to escort a force of Halifax bombers over the target during an attack on an oil plant in the Ruhr as well as their withdrawal. The Manston Wing was also involved in escorting a number of medium bombers that were attacking a petrol and oil dump at Amersfoort, south of the Zuiderzee.

Weather conditions played havoc with operations during this period and on 8th it was noted as being a 'most uninspiring day' with just four operational visitors in the shape of two Typhoons and two Tempests. The 9th was even worse, with just a single Mustang.

Two sections of 229 Squadron were sent to Northolt on the 10th to carry out special escort duties from Northolt, but the weather conditions meant that the operation was cancelled and they returned to Manston. Word came through to Manston that the operations was to be carried out at dawn the next morning

Flight Lieutenant M. Kellet, one of the leading pilots of 504 Squadron, in his Spitfire.

Ground crew of 504 Squadron and the squadron mascot with a Spitfire.

No. 504 Squadron ground crew and pilots at Manston.

and so the section of 229 Squadron Spitfires returned to Northolt so that they could be in position. They returned to Manston the following day after carrying out the special duty, which had meant escorting King George VI's aircraft during his visit to meet General Montgomery and tour units of the 21st Army.

What was described in the ORB as the 'chief point of interest' on the 12th was the emergency landing of a Lancaster from 582 Squadron that was based at Little Staughton. The aircraft had suffered extensive flak damage and both its inner engines were unserviceable, so that it swung to starboard on landing. The undercarriage on the port side collapsed because there were problems with the hydraulics and brakes and the port engine caught fire, but this was quickly brought under control. Rather remarkably, none of the seven-man crew were injured.

There were several emergency landings on the 14th, the first one being a Halifax that had its starboard inner engine unserviceable and extensive flak damage; this was followed by another Halifax with its port inner engine unserviceable and no R/T. Then there was a Liberator with an unserviceable port engine, which was followed by a Flying Fortress that had lost its port engine, but also had its port elevator shot away. A Halifax with its port engine unserviceable completed the immediate emergencies.

That night, six Yorkshire-based Halifaxes landed at Manston in various states of disrepair and damage after a bombing raid on Duisburg; the first two were from 76 Squadron, with engine problems and a shortage of fuel. Another Halifax from 415 (Canadian, Swordfish) Squadron landed with engine trouble and one each of 77 Squadron and 462 Squadron that had been damaged by flak. The pilot of the latter had been wounded while over the target and was in a state of concussion so had to be aided to fly and land the aircraft by the bomb aimer. The rear gunner had baled out over Manston, but landed safely. The final Halifax was from 78 Squadron and it landed with engine trouble.

On 15 October, it was announced in the ORB that Wg Cdr Murray had been promoted to the rank of group captain, so elevating the status of the station. It was also noted that having used the term 'Wing Commander Commanding' for a long while, it would take some time to get used any other.

During the early evening the following day, twelve Spitfires of 229 Squadron were airborne on another special escort for three destroyers of the Royal Navy—on board one them was His Majesty King George VI. He was returning from his visit to 21st Army Group in Holland, where he had been updated by General Montgomery. The King had previously toured the beaches of Normandy and the battlefield at Monty Casino, Italy, and at one point it had been suggested that he was after Montgomery's job. He had replied that this was not the case, but he thought the General might be after his. As a result of the unit providing the royal escorts, it was noted in the Manston ORB that 229 Squadron was now 'muchee swankee'.

Administration Officer Squadron Leader J. J. Secter attended an admin conference at 11 Group HQ on the 17th, and he was not very impressed with the agenda or what those present had to say. He noted in the ORB: 'And they that were wise speak with a loud voice for they not begat much gen'.

At 9 a.m. on the 18th, the Manston Wing, comprising of 118, 124, 229, and 504 Squadrons, was briefed for an operation that involved escort duty in support of a force of Lancasters that were attacking industrial targets around Bonn. The wing was led by Wg Cdr Balmforth and the operation went according to plan. On the same day, 2763 AA Squadron were posted out of Manston and replaced by 2795 AA Squadron.

Wg Cdr Thomas Balmforth, the former Commanding Officer of 124 Squadron, had joined the RAF in October 1938 on a Short Service Commission and in 1940 he had served as a ferry pilot, flying Hurricanes to the Middle East and Malta. He had joined 124 Squadron as a Flight Commander in September 1941, awarded the DFC in May 1942, and taken command of the unit the following month. The wing commander went on to become a regular leader of the Manston Wing.

A number of aircraft from 100 Group operated out of Manston on the 19th, but adverse weather conditions again made things difficult and a Stirling and a Flying Fortress became bogged down on the airfield. The Stirling had collided with another Fortress during its landing run and a number of aircraft had to remain overnight at Manston until the conditions improved.

The Spitfires of 229 Squadron left Manston on the 22nd for Matlask in Norfolk. The unit had only been at Manston for a short time, but it was always a sad occasion when a squadron moved on, especially for those remembering the ones left behind.

There were only a small number of movements the following day, including a section of 118 Squadron that went off on a weather reconnaissance, but had to return early because of the bad conditions. Among the small number of aircraft that landed were two Piper Cubs, which were noted as being 'interesting' in the ORB.

No. 131 Squadron arrived on the 25th for the pilots to be briefed and aircraft refuelled for an operation supporting a force of Halifaxes and Lancasters that were attacking Essen. The Manston Wing was also on escort duties, covering another force of 231 'heavies' attacking Homberg. On the way back, two Spitfires of 504 Squadron collided in cloud some 30 miles from base, although, despite extensive damage to both aircraft, both pilots returned safely to Manston.

It was a bleak day on the 26th, with just four landings all day; among them was a Dakota that had conveyed the AOC of 46 (Transport) Group from Brussels, escorted by two Spitfires. The AOC was Air Commodore L. Darvall, who had been appointed to his post on the 16th, having taken over from Air Vice-Marshal A. L. Fiddamont. There was also a Cessna Crane, a type usually used for pilot training, but on this occasion it was used to transport Colonel Hefley from Advanced Landing Group A.83.

During the afternoon of 28 October, Squadron Leader Hepple, DFC, led the Manston Wing on escort duties for a force of Lancasters and Halifaxes that were attacking Cologne. At 3.36 p.m., a Halifax returning from that operation landed at Manston and it was closely followed by another five Halifaxes and a Lancaster. Two of the Halifaxes had been damaged by flak, while the others had engine trouble. There were 117 landings during the day.

No. 91 (Nigeria) Squadron, under the command of Squadron Leader Bond, was posted into Manston on 29 October. The unit arrived from Biggin Hill and was equipped with the Spitfire IXB. Having been involved in the battle against the menace of the V-1, the Squadron's new role was to provide long-range escort to the bomber force.

On the last day of October 1944, 605 Squadron despatched two Mosquitos on day-ranger sorties to Tutow and Arnhem. On the way back, the aircraft became separated and the Mosquito flown by Flt Lt Craven and FS Woodword failed to return. That night, two more Mosquitos went off on patrols, but both aircraft returned to Manston because of a combination of the weather and mechanical problems.

In October, there were 2,619 landings at Manston, with 227 of them made at night. The ORB noted:

> During the month, operations had developed along fairly regular lines, with the Manston Wing being used whenever the weather was favourable for cover and escort to daylight bomber operations. The Manston squadrons operated in 2, 3 or 4 squadrons giving successful protection to bomber formation on 20 occasions. In addition, almost twice daily weather recces. were flown by the Duty Squadron with frequent shipping recces. to the Dunkirk area.

The usual statistics given in the ORB at the end of each month regarding how many aircraft had landed with battle damage and technical problems and how many had been repaired on the station were no longer published. This may have had something to do with the change of command at Manston and that those figures were no longer appropriate.

On 1 November, fifty-five Spitfires from 26 and 63 Squadrons landed at Manston after 'spotting' for the heavy guns at Walcheren, Holland, for the Royal Navy and flying seventeen sorties. Both were former tactical reconnaissance units, whose pilots had been trained in 'spotting' and ranging the guns for the Navy and they would have only a very short stay at Manston.

Poor weather conditions affected operations again at the beginning of the month, but on the 4th the Manston Wing took part in Ramrod 1359, making sweeps around the city of Cologne, which had been bombed by 180 Lancasters. That night, four Halifaxes and a single Lancaster landed at Manston, with the flight engineer of the latter being very badly wounded.

It was a very busy day on 5 November, when, among other aircraft, thirty-nine B-17 Flying Fortresses were diverted to Manston because of bad weather at their base, Kimbolton. Later on, a B-17 from the 548th Bombardment Squadron and a B-24 Liberator from the 734th at Old Buckenham also made emergency landings.

Wg Cdr Balmforth led 91, 118, and 124 Squadrons on escort duties, providing cover for a force of Lancasters that were attacking Hamburg again. Squadron Leader Bond, the CO of 91 Squadron, had to make an emergency landing at Hawkinge after suffering engine failure. On the same day, Crown Prince Olav of Norway visited RAF Manston; having taken refuge in Britain in 1940, he had been appointed Norway's Chief Defence Minister.

A couple of Vickers Warwicks arrived at Manston on the 10th and later carried out a demonstration in Ramsgate Harbour of its airborne lifeboat that were being used to rescue downed crews from Bomber Command. The lifeboat had been designed and produced by renowned boat designer Uffa Fox, and it was 27 feet long and weighed 1,770 lb—it dropped in the sea with the aid of three large parachutes. Both the public and a number of VIPs watched the demonstration and, according to the ORB, there was 'much reflected glory in which to bask'.

Manston was busy on the 12th, when a variety of aircraft arrived including an Anson, Albemarle, and nine Dakotas carrying civilian passengers. There are no clues as to the identity of the civilians, but the ORB states that they were dealt with by 'continental control'.

During the afternoon of the 14th, a section of 124 Squadron carried out reconnaissance sorties over Dunkirk and reported that there was what appeared to be a submarine tied up to the mole. The following day, there were no operational visitors at all, but 103 Mustangs were diverted to Manston on the 16th, contributing towards the 192 total landings for the day.

Forty-four Spitfires from the Biggin Hill and North Weald Wings arrived at Manston on the 18th to refuel and brief the pilots for an operation to support 479 heavy bombers that were to attack Münster. The raid was led by Wg Cdr Balmforth. On the same day, 605 Squadron despatched two Mosquitos off on day-ranger sorties, but both aircraft returned early because of the weather.

The two Spitfire wings were back at Manston on the 21st to support another raid by heavy bombers on Hamburg. However, it was on this day that Manston said farewell to the airmen and officers of 605 Squadron that had been at the station since April. The unit was moved to Hartsford Bridge, which had been renamed RAF Backbushe just a few days before. The Squadron would continue to operate in the Intruder role until the end of the war and would also continue to operate from Manston.

Later in the day, a Halifax of 425 (Alouette) Squadron from Tholthorpe crash-landed after being attacked by enemy fighters, and the rear gunner was reported to have been seriously wounded. That was followed by a Mosquito from 571 Squadron that also made a crash-landing, but with bombs on board. Its two crew escaped serious injury, but suffered from severe shock and the explosion left a huge hole in the runway that was filled in the next day.

On the day that the runway was repaired, there was a disturbance caused by a number of Naval ratings who were at Manston on 'rest and recuperation'. It was noted in the ORB that those on R&R did not know whether they were 'to be or not to be'—presumably referring to their state of intoxication.

There were a number of Dakotas that required escorts on the 25th and section of 504 Squadron was called upon to fly to Hendon, where it was to join up with the first one and escorted it to Brussels. Another six Spitfires of 124 Squadron flew up to Hendon and met up with another two Dakotas that were flying to Eindhoven. It had been arranged that its escort was to be reinforced by another six Spitfires from 124 Squadron and twelve from

91 Squadron over North Foreland. The latter two units were led by Wg Cdr Balmforth and there are no clues as to who was on board the Dakotas.

No. 406 Squadron arrived at Manston on the 27th, equipped with the Mosquito and posted in from Colerne, Wiltshire. The unit had only recently traded in its Beaufighter VIfs for the Mosquitos and most of its crews were still under training when it arrived.

On the final day of November, fourteen Spitfires from 313 Squadron and another ten from 310 Squadron landed after providing an escort with the Manston Wing for a force of Lancasters that had bombed Osterfield and Bottrop. Flt Lt Chadwick of 504 Squadron experienced engine trouble and was forced to land at Ghent. A Halifax from 192 Squadron made a crash-landing and one of its engines was burnt out, although not due to enemy action. The total number of landings during November comprised:

	Day	Night
Operational Visitors	689	119
Non-Operational Visitors	445	6
Manston Operational Flights	907	66
Manston None Operational Flights	835	22
Total	2,876	213

Final comments in the ORB:

Operational
The Manston Wing carried out many escort operations to bombers attacking targets (mainly in the Ruhr) but no enemy aircraft were sighted. In addition, the Duty squadron carried out early morning weather reconnaissance sorties over the continent almost daily together with frequent anti-shipping reconnaissance sorties to Dunkirk.

The Mosquitos of 605 Squadron had their activities greatly curtailed because of adverse weather at night. There are no claims this month.

General
And so we squelch out way through one of the wettest Novembers ever known, managing to survive, however, and keep 'em flying in spite of the mud menace. As a point of interest to those interested in Medical Research, it is recorded that Wing Commander McDonald, of Headquarters, Fighter Command, was attached to No. 124 Squadron with whom he flew on two or three operational sorties for research purposes.

24

Towards the End

December was another month that began quietly, with only four operational visitors in the form of Mosquitos. Two sections of 504 Squadron carried out a number of weather-reconnaissance sorties, which brought the total number of landings up to seventy-four.

On 3 December, fifteen Spitfires from 340 Squadron arrived from Biggin Hill to support the Manston Wing on an operation providing escorts for a force of Lancaster attacking a dam near Aachen. The operation that had been designated Ramrod 1390 and took off at 8.45 a.m. under the command of Wg Cdr Balmforth. Later in the day, two Mosquitos from 96 Squadron arrived to carry out anti-Heinkel sorties from Manston. This was to intercept Heinkel IIIs that had been adapted to carry a V-I flying bomb, after many of the launch sites on the ground had been overrun by Allied forces.

The Manston Wing encountered large numbers of enemy fighters on the 5th, when they were escorting a force of ninety-three Lancasters that were bombing Hamm. Over 100 Fw 190s and Bf 109s, in two boxes, were spotted by 91 Squadron pilots, who then attacked them head on. Wg Cdr Balmforth spotted that the last Lancaster in the stream was being attacked by a number of enemy fighters and so he led five Spitfires of 504 Squadron to assist its gunners on board the bomber. The Lancaster, with one engine on fire, was escorted to the western side of the Ruhr where it disappeared in cloud, but it must have landed safely because none of the bombers were lost. However, of the Manston Wing, Fg Off. Collier of 91 Squadron was missing, and Fg Off. Topham (from the same unit) made a forced-landing at Gilze-Rijen with engine trouble.

A serious incident was narrowly avoided on 7 December, when, at 5 p.m., a Dakota that was carrying a number of senior officer made an emergency landing because of adverse weather conditions in the Channel. On board the aircraft were the SASO and AOA of the 2nd Tactical Air Force, as well as a number of senior engineering and equipment officers. All those on board were forced to remain at Manston overnight and resume their flight the following morning.

An inspection of the station's catering facilities was carried out on the 7th by Squadron Leader P. S. Goad and J. D. Frame from 11 Group Fighter Command HQ. They professed to be well-pleased with what they had seen and departed in a flurry of good wishes for 'better catering' in 1945.

During the night of the 10th, a Wellington from 24 OTU, based at RAF Honeybourne, Worcestershire, landed at Manston after it had been reported as 'missing' and lost for over three hours. The aircraft had been on a training flight over southern England, but for some reason, probably bad navigation, had crossed the Channel and flown over France, where its crew managed to avoid being hit by flak and attacked by night fighters. The aircraft was 'homed' to Manston by the use of searchlights; when it landed at 1.35 a.m., its fuel had been virtually exhausted.

Another aircraft that was lost found sanctuary at Manston the next night, this time an Albacore from 119 Squadron, attached to 85 Group and based at ALG B.65 (Maldegem) in Belgium. It had taken part in an uneventful anti-shipping patrol when its pilot found himself lost, but managed to cross the Channel and find Manston. The same night, three Mosquitos from 406 Squadron carried out practise Intruder sorties over Holland, but only two returned—Fg Off. Lawless and Fg Off. Reid were listed as missing.

Fog persisted all day on 13 December and there were no operations by Manston-based squadrons. There were only nine landings throughout the day, among which was a Lockheed Hudson that was carrying General Stuart Menzies, who was on his way to London. The aircraft was diverted to Manston and landed with the aid of FIDO, this being only the second occasion when it was mentioned being used. Some sources state that FIDO was used to assist the pilot of the Hudson to land and that is quite likely given who was on board, although it is not mentioned in the ORB.

Among the visiting aircraft the following day was a Dakota that was fitted out as an 'air ambulance' and four Mosquitos of 456 Squadron, which was based at Ford, but regularly used Manston. The Spitfires of 91 Squadron flew a number of weather-reconnaissance sorties and one of them later failed to return to Manston. The Squadron later received the good news from France that Plt Off. Kay had landed safely at Bayeux.

The Spitfires of 118 Squadron left Manston on 15 December for the unit's new base at Bentwaters in Suffolk, where it was to re-equip with the Mustang III. The unit would continue to fly bomber-support escort operations until the end of the war.

On the 16th, the weather interfered with operations again and there was no activity by the Manston-based units, with just twenty-four landings throughout the day. That night, the searchlight cone was lit up for over two hours to assist nine Mosquitos from 627 Squadron at Woodhall Spa and a single Lancaster.

At 6.30 p.m. on the 18th, eleven Halifaxes from 4 and 6 (Canadian) Group began to land after bombing Duisburg, with one them having received a direct hit from flak while over the target. The aircraft had a French crew and was from either 346 or 347 Squadrons, two all-French squadrons based at Elvington near York. Three of the crew had been seriously wounded and another two had baled out over the target area.

That night, sixteen Beaufighters of 16 Group (Reconnaissance) Coastal Command from North Coates landed after flying uneventful patrols. No. 406 Squadron was not yet fully operational and carried out what were described as 'operational training' sorties. There were twenty-six landings during the night.

The Spitfires of 1 Squadron arrived at Manston on 18 December, although they were not mentioned in the ORB until the following day, when the unit was due to fly a bomber-escort operation with 91, 124, and 504 Squadrons. This operation was cancelled because of the weather. The Commanding Officer of 1 Squadron was Squadron Leader D. G. S. Cox, DFC, who had taken over the unit in January.

On the 19th, seventeen Lancasters landed after carrying out operations and one of them had been damaged by flak; however, fog developed after dark and there was no more activity. The following day was much the same, with thick fog. Later on in the afternoon, there was a 'test burn' of the FIDO facility and it noted that blue sky could be seen in places as a result of the burn. Despite the fact that FIDO was being used, there were no landings either during the day or night of 20th and 21st.

On 22 December, there were still no operations flown by the Manston-based squadrons, but there were fifteen operational visitors, including nine Dakotas and two Lancasters. The night of the 22nd was very busy, with one Halifax from each of 4 and 100 Groups and thirty-two Lancasters from 1 Group landing between 7.40 p.m. and 10.30 p.m.

Fg Off. Green of 305 (Wielkopolska) Squadron found himself in difficulties on the night of the 23rd, when his Mosquito was hit and damaged by flak while in the area of Koblenz. There was a loss of hydraulics and the aircraft's brakes, flaps, and navigation instruments were out of action, and so Green was told not to land at the unit's base—Cambrai-Épinoy. They were ordered to divert to the UK and crossed the coast near Newcastle, but instead of being ordered to land at Carnaby or Woodbridge, the other two emergency airfields, they were told to land at Manston.

They were guided by searchlights, which appeared at intervals pointing towards to sky and indicating the route they should fly. They touched down a lot faster than normal, at approximately 150 mph, and they used up a lot of the runway before the Mosquito came to a stop. Green and his navigator were glad to land at Manston, but were puzzled why they had been ordered to fly an additional 200 miles past the emergency runway at Carnaby.

What is very noticeable about the Manston ORB for this period is that it remains 'business-like' and, unlike the previous years, only mentions operational matters. There is no mention of any parties, festivities, or 'jollifications', although they undoubtable still happened at Christmas. Maybe the feeling that the war was so very nearly at an end affected the mood of things or it could have been the scale of the recent German offensive in the Ardennes.

Four Wellingtons from 16 Group (Coastal Command) operated out of Manston on Christmas Eve, and three of them dropped bombs on packs of E-boats that were threatening shipping in the English Channel. No. 16 Group was out in strength and sixteen Beaufighters landed at Manston after carrying out anti-shipping patrols off the Dutch island of Heligoland.

Christmas Day was no different from any other day in 1944, and 1 and 91 Squadrons patrolled Aachen without incident, while 124 and 504 Squadrons were operational over the Bonn–Coblenz area of Germany, where they encountered three German Me 262 jet fighters. The first one was seen patrolling above Bonn and the two others were observed over Duran, but there were no attempts to challenge them. There were ninety-two landings during the day.

Boxing Day was a quiet day because of the weather, it being a very foggy day. Despite that, there were thirty non-operational landings, three of which were assisted with the use of a full burn of FIDO. The airfield was fog-bound again the following day and the only visitor was a B-17 Flying Fortress that made an emergency landing. The aircraft was 42-31484 flown by Lieutenant Stanley Erickson from the 384th Bomb Group, and it came down on the 406 Squadron's dispersal, killing all nine crewmen on board. This has been described as the worst incident to occur at Manston.

The Manston Wing was busy on the 28th, engaged in Ramrod 1419 to escort Lancasters that were bombing the marshalling yards at Cologne. Not all the aircraft returned safely and the Spitfire of Fg Off. Murphy from 124 Squadron was hit by flak and he was forced to bale out over Duren. Later that evening, news came through that he was alive and in an American military hospital in Aachen, suffering from second-degree burns.

There was another disaster that night, when a Lancaster, NN750 UL-M, from 576 Squadron, based at Fiskerton, crashed to the south of Manston village and burst into flames, killing six of the seven airmen on board. The pilot, Fg Off. D. Fletcher, was on his second attempt to land when the aircraft lost airspeed and stalled. The rear gunner, Sgt J. Norris, survived the initial impact, but was seriously injured and died a short while later in the sick quarters at Manston.

Friday 29 December was a very busy day and one of the first aircraft to land was a Halifax from 4 Group, whose navigator had been badly wounded. The aircraft had been escorted to Manston by two Spitfires from 91 Squadron. This was followed by a B-25 Mitchell that had aborted a special bombing operation because of mechanical trouble and another Halifax that had just jettisoned it bombs because it had lost the use of two of its engines. Another two B-17 Flying Fortresses made emergency landings, with both aircraft having problems with their engines.

The Andrew's Field Mustang Wing, based in Essex, were due to land and operate from Manston, but the weather closed in and four of its squadrons were diverted to Bradwell Bay—only 65 Squadron and six Mustangs from 122 Squadron arrived at Manston. The Manston Wing, comprising of 1, 91, and 504 Squadrons, was assigned to Ramrod 1420 to escort a force of Lancasters and Halifaxes that were bombing the railway yards at Cologne.

Fog returned on the 30th and prevented flying operations from taking place and there were just two landings, including a Mustang that needed to refuel. Four Mosquitos of 406 Squadron went off on an operation to patrol enemy airfields, but they failed to find any aircraft and had to be content with attacking a train.

On the last day of 1944, the Manston Wing—1, 91, 124, and 504 Squadrons—took part in Ramrod 1423 to escort a force of Lancasters over the target and cover their withdrawal from the marshalling yards at Solingen. The raid was led by Wg Cdr Bobby Oxspring, who was last mentioned in relation to his part in the Channel Dash in February 1942. It was to be a tragic end of the year for 504 Squadron.

Approximately 40 miles north-east of Brussels, and while flying at 20,000 feet on the way to the target, Flt Lt Fowler's Spitfire was seen to suddenly climb very steeply, turn over on its back, then dive towards the ground. At 10,000 feet one of the aircraft's wings was seen to break off and there was no attempt by the pilot to bale out. There were no enemy aircraft in the area at the time, although it was noted that there was some heavy flak over the target, but what caused the Spitfire to crash remained a mystery. After the operation, 504 Squadron landed at Moorseele in Belgium, but it is not known whether that was because of what had happened to Flt Lt Fowler. Final comments in the ORB:

Summary
The total number of landings during the month of December 1944 was 2,447, some 429 less than in November. There were another 403 landings made at night.

Operations
The Manston Wing continued its role of giving protection to heavy bombers attacking targets, mostly in the Ruhr. It was only very seldom that aircraft were seen. Two enemy aircraft were destroyed by 91 Squadron.

The Duty Squadron carried out almost daily early morning weather recces over the continent.

The Mosquitos of 406 Squadron started to operate during the month over enemy airfields and destroyed two enemy aircraft.

25

The Year of Victory

Ten Liberators and a B-17 Flying Fortress were among the operational visitors on the first day of 1945, which was affected by the weather again. Three squadrons of the Manston Wing—1, 91 and 124—had been briefed at 8.15 a.m. to take part in Ramrod 1423, escorting a force of Lancasters to Ladbergen and the Dortmund Elm Canal near Münster. Unfortunately, half-an-hour before take-off, fog descended on the airfield and they were unable to depart. Also on this day, 1333 Wing of the RAF Regiment was formed at Manston and remained at the station for the duration of the war. The unit was commanded by Wg Cdr T. R. F. Brook.

That night, two Warwicks landed at Manston after successfully deploying two lifeboats during an air-sea rescue mission, in which six airmen were rescued. They were later picked up by a high-speed launch and safely brought ashore. Four 'flower patrols' (patrols of enemy held Dutch airfields) were flown by 406 Squadron, during which they claimed a Me 210 and Ju 88 destroyed.

A B-17, two Liberators, and a Spitfire were among the aircraft that made emergency landings on the 3rd and the Manston Spitfire Wing was briefed to escort a force of Lancasters that were going to bomb Dortmund, but the operation was cancelled. On their return, fifty of the Lancasters were scheduled to land back at Manston and, as the operation was cancelled, it was a quiet night.

There were no operational visitors on the 4th and only twenty non-operational. Pilots of the Manston Wing had been briefed to carry out Ramrod 1426, escorting a force of Lancasters to Ludwigshafen, but the operation was cancelled because of the weather. A section of 504 Squadron carried out weather-reconnaissance sorties to Amsterdam, Eindhoven, and Amiens. The Ludwigshafen raid took place on the 5th, with the Manston Wing providing the escort on what became Ramrod 1427.

An Airspeed Oxford that was flying from ALG B.83 (Knocke le Zout) in Holland to Bentwaters crashed approximately 5 miles from Manston on the

6th. The pilot and three of the passengers were taken to the sick quarters and there is no further information about how many people were on board or who they were.

Snow began to fall after dark on the 7th, and a gale warning was issued for south-east England during the early hours of the following morning. Despite the bad weather, two Liberators and four Lancasters landed with various states of battle damage and engine trouble. Three Mosquitos from 406 Squadron carried out practice flights, while another aircraft carried out an Intruder sortie over Germany.

More snow fell over the night of the 8th into the early hours of the 9th, when a section of 504 Squadron was scrambled to intercept an unidentified aircraft that turned out to be friendly. There was a considerable fall of snow on the 10th, which caused the 'Manston Snow Plan' to be put into operation. The Snow Plan involved lots of men clearing the active area with shovels. On grass runways prior to 1943, there was always a danger that the grass surface could be damaged and turned into mud, and so it was often left to thaw and clear itself. With the advent of concrete and asphalt runways, it was much easier and often involved mechanical means, such as the 'snow blow'—this was a jet engine.

The Manston Wing was active of the 11th and was called upon to escort a force of 150 Lancasters that were to bomb the railway centre at Krefeld, but the aircraft were recalled while in the area of Schouwen-Duiveland.

No less than sixty-four B-17 Flying Fortresses and sixty-seven P-51 Mustangs were diverted to Manston on the Thursday 13 January, with the crews being accommodated overnight. With each B-17 having a ten-man crew, together with the pilots of the Mustangs, this meant food and accommodation had to be found for over 700 airmen. The Manston Wing was briefed to take part in Ramrod 1432, but, half-an-hour before take-off, the operation was cancelled.

The following day, forty-three Piper Cubs arrived at Manston to be refuelled before flying to the 'other side'. With its 65-hp Lycoming engine and a cursing speed of 75 mph, the Piper Cub could hardly be described as a war plane, but it did play its part in the conflict, mainly being used in the reconnaissance and transport role. A small number of them were fitted with racks under their wings, on which bazookas were fitted to be used against enemy armour. It has been claimed that on one occasion a Piper Cub destroyed six enemy tanks, but this cannot be confirmed.

Nos 6006 and 6616 Servicing Wings left Manston on the 14th, along with 1401 Meteorological Flight, which had been based on the airfield since September 1943, but rarely mentioned in the ORB. No. 1007 Servicing Wing moved in to Manston the following day.

The Meteors of 616 Squadron left Manston on the 17th for their new base at Colerne in Wiltshire. Despite the fact that the Squadron had been at

Manston since 21 July 1944, their operations had rarely been mentioned in the ORB.

The station motor transport officer, Flt Lt Murton-Neale, took over from Flt Lt L. C. Carter on the 19th. That night, two Wellingtons from the 2nd Tactical Air Force landed at Manston after carrying out reconnaissance sorties over the Ruhr Valley. There was no activity among the Manston-based squadrons.

Escort duties were not always for heavy bombers and, on the 21st, 91 Squadron escorted a Lockheed to Brussels; the following day, 124 Squadron carried out escort duties for a Lockheed to Ghent. There is no mention of what type of Lockheed these aircraft were, but the chances are that they were Lockheed Lodestars, a twin-engined passenger aircraft that was often used to ferry senior officers and politicians to important meetings.

The Under Secretary of State for Air, Captain Harold Balfour, landed at Manston on the 23rd on his way to London. The ORB does not tell us where he flew from or what type of aircraft he was in, but it was probably a Lockheed Lodestar.

Extremely bad weather clamped down on the airfield on the 26th and there was only one non-operational landing in the form of an Airspeed Oxford. The following day, it was still very foggy and the only landing was made by a solitary Spitfire.

The Manston Wing was briefed to take part in Ramrod 1455, but its part in the operation was cancelled just before take-off. Five Mosquitos of 128 Squadron and seven from 162 Squadron were diverted to Manston, but otherwise things were generally quiet because of the weather. On the final day of January, a Mosquito and an Anson were the only movements and there was no other activity. Final comments in the ORB:

Summary and Conclusion
Total number of landings in January—2,137, of which 246 were by night.

Owing to unfavourable weather conditions, operational activity was greatly restricted. The Duty Squadron continued early morning weather recces over the continent and the Spitfire Wing carried out a few escorts to bombers attacking targets in the Ruhr area.

Mosquitos of 406 Squadron continued to operate over enemy airfields and destroyed 1 Me 110, 2 JU 188, and probably 1 Ju 188 at night.

On 1 February, the Manston Wing was involved with Ramrod 1448, covering the flanks of the bomber stream that was attacking Mönchengladbach. That night, a Lancaster from 5 Group landed on only two engines, having sustained flak damage over Seigen.

During the night of the 2nd, a Halifax (NP819 OW-B) from 426 (Thunderbird) Squadron, based at Linton-on Ouse, crashed while returning from a bombing sortie at the oil refinery at Wanne-Eickel. The incident happened after the pilot, twenty-six-year-old Fg Off. Joseph Peter Talocka, had been unhappy with his first approach and had flown a wide circuit of the airfield before losing control at 600 feet. All attempts by the police, the NFS, and the RAF's own ambulance party failed to find the aircraft, but one surviving member of the crew, Canadian Warrant Officer McAllister, who had baled out, thought that the aircraft had crashed in the area of the Minster Marshes.

The aircraft was not found until the following day, when the wreckage was discovered close to Castle Farm near the historic remains of Roman-built Richborough Castle, between Minster and Sandwich—all other members of the crew were dead. The only British member of the crew was thirty-two-year-old Sgt Graham Needham from Scunthorpe—his body was returned for burial. The bodies of four of the Canadians in the crew, including that of the pilot, were buried at Brookwood Cemetery in Surrey (London Necropolis). The others were twenty-year-old Plt Off. Jack Morris Styles; twenty-four-year-old Sidney George Arloote; and twenty-year-old FS Joseph Alexander Chisamore. It is not known where the fifth crew member, FS Alan George Bradley, was buried.

No. 426 Squadron was part of 6 (Canadian) Group, based in Allerton Castle, near York, where those Canadians from the RCAF who served in the Second World War are remembered. On Saturday 6 September 2014, there was a remembrance service and parade at Allerton Castle to commemorate Canadian aircrew and celebrate the fact that a Canadian Lancaster had flown over from Canada and was touring Britain. Senior officers from both RAF Linton-on-Ouse and RAF Leeming were in attendance and there was a flypast by both the Canadian Lancaster and PA474 from the RAF's Battle of Britain Memorial Flight.

A section of 91 Squadron took off on weather-reconnaissance sorties on 3 February, and another section returned from Northolt after escorting a Dakota from there to Brussels. The Manston Wing took off at 2.15 p.m., led by Wg Cdr Oxspring, to carry out Rodeo 411, a sweep of airfields in the area of the Rhine, and afterwards landed at Ursel as ordered. The number '411' seems to be out of sequence with other sorties flown and we can only presume it to be an error. Another Lancaster from 582 Squadron landed that evening after aborting an operation to Bottrop because of engine trouble. Those sorties when a pilot turned back because of mechanical trouble were known in Bomber Command as 'boomerangs' and both pilots and crew often came under scrutiny and the aircraft thoroughly checked when they returned early.

The Manston Wing returned from Ursel on the 5th; on this night, a VIP was due to arrive, but for some reason his aircraft was diverted to Tangmere. The next day was similar to that of the previous one and the Manston Wing took off on Ramrod 1453 to Paderborn and then landed at Ursel to refuel before flying back to Manston again. No. 124 Squadron flew off to Swannington, Norfolk, where they were to be bombed up and refilled for bombing sorties on objectives in Holland. However, when they arrived at Swannington, they were told the operation was cancelled and so they had to return to Manston.

Two more Halifax from 6 Group landed during the early hours of the 9th after another bombing raid on Wanne-Eickel; both aircraft had been damaged and lost an engine. Manston was becoming a regular sanctuary for the Canadians. The Spitfire Wing led by Squadron Leader Cox went off on Rodeo 414, another operation that was seemingly out of sequence. The Wing flew to Ursel where they refilled before taking part in sweeps of the Rhine and Osnabrück area.

Among the visiting aircraft on the 11th were twenty Dakota C.47s, which were carrying wounded American servicemen. They were diverted to Manston because all the American airfields were closed because of the weather. The station hospital at Manston had only thirty beds and it was already busy—they were given just twenty-five minutes' notice to make arrangements for 175 patients.

The operation to deal with such a high number of casualties involved the ARP ambulance crews and other civilian organisations, which were called upon to transport wounded airmen to Margate hospital. The following day, the operation was to have been reversed with the wounded being flown out to American bases, but the weather clamped in again and the Dakotas were forced to return to Manston. Eventually, a hospital train was organised and the wounded men had to be transported by rail.

The night of 14 February proved to be both busy and eventful and the first aircraft to arrive was a Lancaster from 1 Group that had engine trouble and five Halifaxes from 4 Group. One of them had an engine unserviceable, while another had been badly damaged by an enemy fighter. They were followed by two Lancasters from 5 Group, one that had been damaged by flak and another by fighters. A Mosquito from 8 (Pathfinder) Group landed short of fuel, along with a B-17 from 100 Group that had two wounded on board. No. 406 Squadron were busy and flew ten operational sorties, with Flt Lt Groome and Fg Off. Johnston claiming a He 219, twin-engined night fighter near Erfurt.

Bad weather prevented flying on the 15th and the ORB stated: 'No activity is reported by this station'. The Manston Wing remained at Ursel and did not return to Manston until the afternoon of the 17th. On the night of the 18th, 406 Squadron flew a number of high-level patrols over Germany and one crew (that of Fg Off. Hamlyn-Love and Flt Lt Radcliffe) failed to return.

During the night of the 21st, three Lancasters landed, one of which had sustained some very serious damage and caught fire as it touched down.

Fortunately, most of the crew were able to evacuate the aircraft without injury, with the exception of the rear gunner, who fell out of his turret and injured his head.

Manston was 'invaded' the following day by forty Spitfires of the Bradwell Bay Wing, comprising of 310, 311, and 313 (Czech) Squadrons that had escorted eighty Lancasters during a raid on Oberhausen. Meanwhile, the Manston Wing, under the command of Wg Cdr Oxspring, took off from Ursel to escort a force of Lancasters attacking Gelsenkirchen and then returned to Manston in the afternoon.

No. 504 Squadron left Manston on the 25th for its new base down the coast at Hawkinge, where it would remain for just over a month. In April, 514 Squadron would move again to Colerne, Wiltshire, to enter the jet age and re-equip with the Meteor III. What the modern jet fighter was capable of was witnessed by a couple of pilots from 91 Squadron, who saw a Me 262 attack a Lancaster during Ramrod 1479, a bombing raid on Cologne. They saw it attack the Lancaster, sending it down in a spiralling ball of flames, but the Me 262 moved so swiftly that there was no way they could have engaged it.

The Me 262 was vulnerable to piston-engined fighters and, on 5 October 1944, a number of Spitfire IXs from 401 (Canadian) Squadron, under the command of Squadron Leader Rod Smith, had destroyed the first one. Their victim was *Hauptmann* Hans-Christof Buttmman, a bomber pilot from KG 5. According to Adolf Galland, bomber pilots flying fighters was a hopeless cause and many others were killed in a similar fashion.

On the 27th, the three Czech squadrons were back and posted to Manston, a move that was described as the 'Czech invasion'. The units, led by Wg Cdr Hlado, landed during the early evening after escorting a force of bombers to Gelsenkirchen. At Manston, the Bradwell Bay Wing became the Czech Wing and it was noted in the ORB that both officers and airmen wasted no time in settling down.

The Manston Wing lost an aircraft on the final day of February, when Flt Lt Draper made a crash-landing shortly after taking off from Maldegem in Belgium. It was later announced that Flt Lt Draper was safe, although his aircraft had been written-off. Final comments in the ORB:

Summary and Conclusion
During February, there were 2,547 landings, 341 of which were by night.
　The unfavourable weather again restricted operations, but the Spitfire IXs (1, 91, and 504 Squadrons) continued the role as supporting heavies attacking targets, mostly in the Ruhr.

On 1 March, twelve Spitfires from 1 Squadron flew to Northolt to be in position to carry out escort duties the following morning. At 11 a.m. the following morning, they took off to escort an aircraft carrying a passenger described as a 'very, very important person (VVIP)' to Brussels. On their return to Manston, Squadron Leader Cox, the CO of 1 Squadron, crashed his aircraft and it was written-off, although he was lucky to escape serious injury. The 'VVIP' was almost certainly Prime Minister Winston Churchill, who is known to have toured the battlefields of Europe between the 2nd and 6th March.

A B-24 Liberator (44-40436, *Nancy*) diverted to Manston on Friday 2 March after it had been hit by flak over Calais. Waist gunner S/Sgt E. A. Cinquina was wounded in this incident. The aircraft was flying at 4,500 feet when a fragment of flak struck his jugular vein and, despite all the efforts to save him, he died before the aircraft could reach Manston.

No. 406 Squadron carried out seven Intruder sorties on the 3rd and another two Mosquitos were scrambled against incoming enemy aircraft, but one of them was then ordered to carry out a ranger sortie. The Mosquito that went off to intercept the enemy aircraft, flown by Flt Lt Donovan and Fg Off. Grant, failed to return and was last heard 30 miles from base over the North Sea.

A new record number of seventy landings was made at night on the 5th after a raid by Bomber Command on Chemnitz, involving 498 Lancasters and 256 Halifaxes. Thirty-five Halifaxes from 4 Group and five from 6 Group were among those that landed at Manston, along with three Lancasters from various other groups. Icy conditions had badly affected the operation and nine Halifaxes from 6 Group crashed shortly after taking off from their bases in Yorkshire. No. 426 Squadron, which had recently lost a Halifax near Manston, lost another three out of its fourteen aircraft soon after taking off from Linton-on-Ouse because of the icy conditions.

On the night of the 7th, Mosquitos of 406 Squadron carried out six Intruder sorties, but Fg Off. Oswald and Plt Off. Hicks failed to return. Nos 1 and 91 Squadrons were taking part in Ramrod 1487 on the 9th, and afterwards landed at Maldegem. Fg Off. Hyde of 91 Squadron was acting as a relay for R/T messages on the operation, but his signal became weak and he asked for a 'homing' back to Manston. Hyde was told to climb to give a stronger R/T signal, but nothing was heard from him again and he was listed as missing. Three days later, news was received from France that Hyde was safe.

Flt Lt R. S. Sowden arrived from Hunsdon on the 10th to take over the post of station adjutant. He replaced Flt Lt G. R. Miller, who had held the post for eighteen months and had been posted to Biggin Hill to take over as sector adjutant.

Ramrod 1493 took place on the 13th, involving both the Czech Wing, led by Wg Cdr Hlado, and the Manston Wing, led by Wg Cdr Oxspring. The squadrons took off from Ursel and Maldegem respectfully, and afterwards the Czech Wing returned to Ursel, while 1 and 91 Squadrons landed at Manston after an uneventful operation.

A Spitfire from 1 Squadron was scrambled on the 16th to assist the pilot of a Mosquito that was in trouble, and the aircraft crash-landed at Manston. Another Spitfire was scrambled during an 'X' Raid (unidentified aircraft), but it turned out to be a Meteor, a type of aircraft rarely mentioned in the records.

On the night of the 19th, gale-force winds from the south-west (with gust in excess of 50 mph) interfered with flying operations, but, 2 a.m., two Mosquitos from 406 Squadron were scrambled on a ground-control interception (GCI radar) alert. The alert proved to be a false alarm and they returned to Manston after a short, fruitless patrol.

The following night, there was an air-sea rescue alert and a Warwick from 280 Squadron, based at Limavady in Londonderry, landed after co-ordinating a rescue. While carrying out a patrol of the North Sea, the crew of the Warwick had come across a Walrus that was being swamped by heavy seas and its crew had been taken off by the crew of Catalina flying boat. There was little the crew of the Warwick could do, but remain on station and maintain communications until the ASR launches arrived.

The Czech Wing, along with 91 Squadron, carried out Ramrod 1513 on the 23rd, escorting a force of 100 Lancaster that were bombing Wesel. After the operation, 310 and 312 Squadrons landed at Maldegem. No. 91 Squadron lander at Ursel, while 313 Squadron returned to Manston. No. 1 Squadron, which had flown to Northolt the previous day, also landed at Ursel after escorting a transport aircraft.

The crew in one of the Lancasters on the Wesel raid from 5 Group got into difficulties after a bird strike that shattered the Perspex windscreen shortly after it had left the target area. The pilot was temporarily blinded and his eyesight badly affected, but, despite that and some damage to the front of the aircraft, with the help of the crew he managed to reach Manston. He then made what was described as a 'rather heavy' landing, which damaged the aircraft, but all the other members of the crew escaped injury.

26

The Final Chapter

Operation Plunder (the crossing of the Rhine by troops of the British 2nd Army) and Operation Varsity (the airborne assault) had begun on the night of 23 March. The following day, a number of aircraft and gliders that had been involved landed at Manston. The first to land was a Halifax that was towing a Hamilcar glider, which had suffered mechanical problems on its way to the continent and had remained connected to its tug.

This was followed by another two Halifaxes from 38 Group that had acted as tugs and had released their gliders as planned. Next were three American Liberators that had dropped supplies to the airborne troops on the ground; these were followed by another one, which had been badly damaged by enemy action.

Damage to 42-51504 from the 714th Bomb Squadron was so extensive that the pilot was unable to land the Liberator and the crew abandoned it, baling out over Manston. The sortie had begun at 9.30 a.m., and the Liberator, flown by Lieutenant Voigt, had taken off from Seething on what the crew had assumed to be a 'milk run', dropping supplies to troops that were about to cross the Rhine. With the exception of just a few minutes at the end of the sortie, they had been over friendly territory all the way, but that was enough time to be targeted by the German guns.

Everything went as planned and the supplies were dropped on the target area, which had been marked by British troops forming a circle of gliders. However, it was at this point that 'all hell broke loose'. Bullets began to rake the floor of the aircraft, narrowly missing members of the crew, but puncturing the nose wheel tyre and shooting away the controls. The Liberator was at 50 feet, with the crew unable to control it; however, the one thing that saved them was the automatic pilot, and when Lieutenant Voigt switched it on, he was amazed to find that it continued to work.

Slowly climbing to 7,000 feet, they made it to the coast at Ostend, where they decided to make for Manston. Over the airfield, they circled for half-an-hour and an assessment of the situation found that the ailerons and rudder were useless. The only control Voigt had was with the elevators—to take the aircraft either up and down—and he was able to bank to port but not to starboard. When the crew baled out, the Liberator was set on a course to crash out at sea in the Channel.

Navigator Lieutenant Harold Dorfman was the second man out and despite the fact that his parachute was soaked with hydraulic oil, it opened and he drifted gently down to earth. As he approached the ground, he noticed a group of women running towards him from a church, but he soon realised that they were not there to welcome him but to get his nylon parachute. The situation was such that it quickly got out of control and Dorfman had to pull out his gun to hold them until an RAF officer arrived. He tried to explain that if he failed to return the parachute he would have to pay for it (approximately $300). After the RAF officer had arrived, he showed the women that his gun was not actually loaded and they became really angry.

Dorfman managed to obtain transport, which took him to Margate where he had tea and was soon reunited with Lieutenant Voigt and the rest of his crew at Manston. Colonel Heber Thompson, who been acting as co-pilot, was rather annoyed because he had landed on his face in a field full of freshly fertilised manor. Dorfman, Voigt, and their crew met up with their CO, Colonel Westover, who had also made a forced-landing with his aircraft at Manston after a couple of his crew had been wounded. Eventually, two other B-24s arrived to return the crews to Seething, but the one Voight and Dorfman boarded had technical problems with its undercarriage that initially refused to lower and they were advised they might have to bale out again.

After twenty minutes circling the airfield at Seething, the wheels were finally lowered for a safe landing, and by 7 p.m. they were safe and sound on the ground—or so it initially seemed. Dorfman returned to the flight line to take some photos of the aircraft and unfortunately the hand brake on the front wheel of his bicycle jammed and he went 'flying' over the handlebars. As a result, he and ended up in hospital with other members of his crew who had been wounded. They had all qualified for the Purple Heart, but all he had to show for his exploits was a busted bicycle.

There were just five operational visitors on the 26th, made up a Boston and four Flying Fortresses, which diverted to Manston because of either engine trouble, fuel shortage, or flak damage. The rear gunner of the last Fortress to land was badly wounded and taken to the station's sick quarters. There was no further activity by the Manston Wing because of bad weather.

Four Flying Fortresses were the only operational visitors on the 28th and there was no flying by the Manston Wing because of adverse weather conditions. That night, bad weather again curtailed all activity. There were few movements the following day either, although 91 Squadron scrambled a Spitfire to intercept a V-1 and patrol the Thames Estuary, but nothing was seen.

Seven Spitfires of 91 Squadron positioned to Northolt on the 31st for escort duty the following morning. Some familiar faces appeared as well when the Meteors of 616 Squadron passed through Manston on the way to its new base at Gilze-Rijen in Holland. Final comments in the ORB:

Summary and Conclusion

There were 2,949 landings at this station in March, of which 342 were made at night.

The Spitfire IX Wings, Nos 1 and 91 Squadrons, and the Czech Wing (310, 312, and 313 Squadrons) continued their role of giving protection to heavies bombing targets in Germany, with an occasional 'sweep', without, however, making contact with any enemy aircraft.

Early morning weather recces over the continent were also carried out almost daily, as well as escorts to aircraft carrying VIPs.

The ORB for 1 April stated: 'A very quiet commencement to the month—no operational visitors'. No. 91 Squadron had sent six Spitfires to Northolt for escort duty, but, although this was cancelled, the aircraft remained there. It was much the same the following day, with no operational visitors.

On the 3rd, the Czech Wing, along with 1 and 91 Squadrons, returned from Maldegem after carrying out Ramrod 1528, providing top cover for 250 Lancaster attacking Nordhausen. The night of the 4th was busy, and three Lancasters from 3 Group diverted to Manston, one of them having been hit by flak with three of its crew wounded. Another member of the crew had fallen out of the aircraft while waiting for the order to bale out and there is no mention of whether it was deliberate or if the had his parachute on.

Two Spitfires from 91 Squadron were called upon to escort a Dakota that was carrying VIP passengers to Gilze-Rijen in Holland on the 6th. Additionally, a Mosquito made an emergency landing on a single engine; the aircraft did not belong to the RAF, but was in the service of USAAF. The 653rd Bomb Squadron operated Mosquitos from Watton, flying high and fast in the light-reconnaissance role, mainly unarmed.

A ceremonial parade was held at Manston during the morning of the 7th, when five Czech airmen were awarded and presented with their DFC medals by the AOC of 11 Group, Air-Vice Marshal John Berresford Cole-Hamilton. Several Czech dignitaries were present, including General Janousek and Group Captain Mrazek. Just four months later, while he was still commanding 11 Group, the AOC died of cancer, shortly after his wife had died.

Later that day, five Swordfish from 157 Wing, 16 Group, were back at Manston to carry out patrols in the Channel. The aircraft belonged to 119 Squadron, which was based at Bircham Newton, but were they temporarily based in Belgium. The Swordfish flew their first patrols the next day and found a number of mines floating loose, endangering shipping. What they were probably looking for was the German *Seehund* midget submarines that had been despatched into the Channel to disrupt convoys carrying supplies to France.

Manston lost two units of the Manston Wing on the 8th, when 1 and 91 Squadrons were posted to Coltishall and Ludham respectively. Both units had spent nearly six months at Manston and their departure spelt the beginning of the end for the Manston Wing.

A Lancaster from 550 Squadron, based at North Killingholme, landed at Manston on the 10th on just two engines, a brilliant display of airmanship despite the fact that the aircraft had no bombs on board. No. 406 Squadron were active supporting a bombing raid on Leipzig, and Flt Lts Etienne and Boaks claimed to have destroyed a He 111 and a Ju 88.

It was also on the 10th that former Manston Commanding Officer, Group Captain Gordon Learmouth Raphael, DSO, DFF, was killed when the Spitfire he was flying collided with a Dakota in the air near Ashford. The twenty-eight-year-old officer had only recently married and was living in Bournemouth with his wife Dorothy. The accident happened in broad daylight and clear skies; there is no explanation as to why he, and the pilots of the other aircraft, failed to see each other. Group Captain Raphael was buried in Cudham's St Peter & Paul's churchyard, Orpington.

Another nine Dakotas landed on the 11th, along with six Spitfires of 313 Squadron, which, having taken part in a bombing raid on Nuremburg, returned to land at Manston—the pilots having ignored orders to land at the ALG B.90, Kleine Brogel in Belgium. The reason the pilots gave was that the airfield was overcrowded, but, after landing at Manston, orders were received from Biggin Hill that they should return to B.90 immediately. Only four Spitfires took off to return to Belgium and just before they reached the Belgian coast, orders were received from the same authority at Biggin Hill for them to return to Manston.

Escort duties continued to be a major role for the fighters and, on the 14th, two Spitfires of 310 Squadron escorted a Dakota to Krefeld Uerdingen without incident. Three Fairey Barracudas, a type of aircraft new to Manston, landed at the airfield on the same day to carry out anti-submarine patrols in the English Channel, under the command of Lieutenant Commander Davies. The strange-looking aircraft that were torpedo bombers belonged to 822 (FAA) and the unit had only recently returned to the UK on HMS *Rajah* after serving in India. By the 16th, another three Barracuda TRIIIs had arrived at Manston and a total of six of them carried out anti-submarine patrols at dawn that day, suggesting that there was still a threat from the Kriegsmarine's midget submarines. Earlier in the year, the midget submarines had operated off the coast of Margate and that may have been the reason why the patrols were carried out.

A Seafire was among the operational visitors on the 19th, and 822 Squadron continued to search the Channel for signs of the *Seehunds*, but all the patrols were uneventful. The following day, five Spitfires and five Typhoons arrived at Manston from 84 Ground Support Unit, which was a support organisation of the 2nd Tactical Air Force that supplied both aircraft and pilots for those lost in action. They had just carried out a successful attack on a pumping station at Dunkirk using rocket projectiles and bombs.

On the 21st, the Barracudas of 822 Squadron carried out three patrols in the morning and another three in the evening and although its patrols proved uneventful the midget submarines were being found and destroyed. On the night of 11–12 March, two of them had been sunk by a Swordfish from 157 Wing and another by a Spitfire. During the 23–24 March, the Royal Navy frigate HMS *Retalick* sank four of them.

From the middle of April, the Mosquitos of 29 Squadron, which was based at Colerne, began to support operations carried out by 406 Squadron. On the night of the 24th, 406 Squadron lost another when Mosquito call sign *Verdict 16*, crewed by Plt Off. Norman and navigator Fg Off. Warwick, failed to return.

The next day, 310 and 313 Squadrons took off for ALG B.86 (Helmond) to take part in Ramrod 1555, supporting a force of 446 heavy bombers attacking

Two Fairey Barracudas in formation—a strange-looking aircraft that arrived at Manston in early 1945 with 822 (FAA) Squadron on anti-submarine duty.

the German Air Force base at Wangerooge. After the operation, the Spitfires returned to B.86, where they spent the night, with the exception of a single aircraft that landed at Manston because of technical issues.

The two Czech squadrons returned from B.86 on 26 April. There are few other details available for the next few days, except to mention that there were ten non-operational visitors on the 27th. No. 822 Squadron also carried out eight uneventful patrols and forty-four non-operational visitors arrived on the 28th.

Another Mosquito from 29 Squadron was lost during the night of the 29th and its crew, Flt Lts Bennett and Oxborrow, were presumed to have been killed after it had been hit by flak near Dunkirk. According to the ORB, it was not on an operational sortie but a night-flying test (NFT). The incident was witnessed by the crew of another Mosquito, which was also hit and damaged by flak, although its pilot managed to land safely.

There were ten non-operational visitors on the last day of April, with 822 Squadron continuing to fly early morning anti-submarine patrols that proved to be uneventful. Among the small number of aircraft at Manston was a Hurricane of the Air Despatch Letter Service (ADLS), which arrived from Northolt. This aircraft was almost certainly from 1697 ADLS, based at Northolt, and had been formed after D-Day to convey secret mail, documents, and equipment to the continent. The Hurricanes had been specially adapted with under-wing storage areas and another secret compartment under the nose.

Summary and Conclusion

	By Day	By Night
Total number of operational visitors	187	46
Total number of non-operational visitors	382	17
Manston operational flights	343	136
Manston non-operational flights	1,091	152
Total number of landings in April	2,003	351

The Spitfire IXs of 1, 91, 310, 312, and 313 Squadrons continued the role of giving protection to RAF Lancaster and Halifaxes bombing targets in German, but the enemy aircraft opposition was conspicuous by its absence. Several escorts to VIPs were also carried out as well as early morning weather recces over the continent. Nos 1 and 91 Squadrons left Manston for Coltishall sector on 9 April. In the middle of the month, 822 (FAA) Squadron, operating under Coastal Command, arrived and were engaged daily in patrolling the Channel searching for midget submarines. None, however, were sighted.

The Mosquitos of 406 and 29 Squadrons, the latter having arrived at the beginning of the month, carried out night patrols over German a/Fs [airfields] and also several rangers. No. 406 Squadron was particularly successful and destroyed 10 enemy aircraft in the air and 8 on the ground and damaged 16. No. 29 Squadron's score was 3 destroyed on the ground and 4 damaged.

An educational vocational training committee was set up on 1 May to make suggestions and proposals for retraining service personnel after the war. Group Captain Murray was chairman of the committee along with Chief Technical Officer Wing Commander A. W. Tillman; Station Intelligence Officer Squadron Leader J. H. Baldwin; Assistant Intelligence Officer Squadron Leader Riley; Education Officer J. O. D. Driscoll; and Senior WAAF 'G' Officer Flight Officer L. J. Snow. Suggestions were put forward and proposals accepted and rejected with equal rapidity, but, after ninety minutes, what was considered to be a safe plan had emerged.

The next day, the education hut was the scene of some heated discussion, when forty officers were led by the Station Commander to discuss the 'discussion groups'. This was in connection with the future resettlement and training of personnel based at Manston, and the Station Commander outlined to the 'silent multitude' how every officer could play an important part in the planning and operation of discussion groups. At the conclusion of his talk, widespread enthusiasm had become most apparent and each officer leaving the education hut had resolved to do the best for all the gentlemen in their sections.

While the discussions were going on, Sgt Kratochive of 310 Squadron took off on a cross-country sortie, but he had to make a forced-landing near Haslemere, although he escaped with minor injuries. No. 406 Squadron flew two successful sorties to Denmark and Fg Off. Kay and Flt Lt Tindell destroyed two Ju 52s on the ground at Marraeback. Fg Off. Greene and Flt Lt Wayman destroyed three at the same airfield.

The ORB for 3 May stated:

> Today marks a great event of a full FIDO burn witnessed by the following very eminent gentlemen: Major General Donal Banks, CB, DSO, MC (Director of Petroleum Warfare); Colonel Lord Suirdale and Sir William Wiseman (Petroleum Warfare Department); Air Marshal Johnson (Air Officer Commanding RCAF, Great Britain); Mr W. P. Harley (Deputy Director, Flying Control, Air Ministry); Group Captain N. W. S. Robinson, CBE (Sector Control, Biggin Hill); Wing Commander Ingall (Petroleum Warfare Department); and Mr Moor (Petroleum Warfare Department, Designer of Hades equipment).

The burn was carried out according to plan and a subsequent report was written up by Flt Lt G. Field—officer in command of FIDO system Manston, and it was not good news. It mentioned a number of potential problems, including the fact that the feed pipe to the igniter was too close to the burn line and it had bulked up above the ground. Also the pre-caste concrete trench of the vaporizer unit was showing signs of disintegration and the establishment of the pump house needed to be increased by two airmen.

According to the ORB 'weird and wonderful' rumours abounded about post-war station activity as the cessation of hostilities became imminent and the Station Commander was of the opinion that the best plan was to provide 'Pukha Gen'.

There was very little to report and very few aircraft movements on the 5th, 6th, and 7th. The great question on everybody's lips was 'is it or ain't it?' the 'it' being the cessation of hostilities with German. Everyone suddenly had a burning desire to be close to a radio and the question was always 'has the Prime Minister spoken yet?'

The station administrative officer, surrounded by a great pile of papers, signals, and maps of the camp, got together with the station warrant officer and they decided on two things immediately: a victory parade with 100 per cent turn out on VE Day; and a grand victory dance on VE Day. The only activity reported on the 7th were weather-reconnaissance sorties by 313 Squadron and four uneventful Channel patrols by 822 Squadron.

The victory dance was discussed at length at the Station Commander's weekly conference and it was decided to hold it in the hangar on the loop (that is still there) that was used by 43 (Maintenance Command) Group. Having agreed on this, the great chase for a sufficient supply of beer began—described as the great beer chase.

'VICTORY IN EUROPE DAY': the words appeared in the ORB in capital letters and expressed great emotional turmoil. The author went on to say:

> At this juncture, our spirits (all types) overflow so much that we could cheerfully indulge in a pointless effusion regarding the termination of a bitter struggle against overwhelming odds. We will, however, as custom wills, cheep feebly, 'Oh, good show'—and give a brief account of Manston activity on this day of days.

At 11 a.m., there was a parade of all station personnel and an address by the Station Commander. In spite of several days of anticipation, many found it difficult to realise that they were being told that the 'great job' was over and that they may now be stood down. After the Station Commander's address, there was a brief service, after which personnel were free to go their own way or return to their duties until the dance in the hangar on the loop.

According to the ORB, the dance went off with a 'bang':

> Everyone was sublimely convivial, that is with the exception of a sad looking 'erk' in the corner, who moaned continually about his post-war prospects. However, after a few of the necessary, he is quite ready to agree that he will in all probability be left a fortune—the war has been won and Oh Boy, is he happy!

It was claimed that the band, conjured up out of the station administrative officer's field service cap at the last minute, had performed wonders and that sentiment was endorsed by one airman, who danced by himself the whole evening. Thoughts were going out for two officers who missed the celebrations—Flt Lt K. D. Hayward and Fg Off. J. M. Smart were ordered to report to Fighter Command in preparation for overseas postings. Intelligence Officer Flt Lt Knight was also posted out to Hutton Cranswick and many others would soon be posted.

Over the next few days, things slowly normalised and, on VE Day+1, 29 and 406 Squadrons each sent six Mosquitos to the Channel Islands, where a landing party was taking control. The sight of the Mosquitos was greatly appreciated and the bailiff of Jersey sent a message of thanks to RAF Manston.

Captain S. Velebny and the 16th Mobile Reclamation and Repair Squadron left Manston on the 12th, and, in a letter to Group Captain Murray, he expressed the deepest appreciation for the splendid co-operation and help given at all times. RAF Manston had answered many prayers offered up by exhausted aircrew and Squadron Leader Ian MacGregor, Commanding Officer of the hospital, was later awarded the Bronze Star Medal by the Americans for saving the lives of many of their wounded aircrew. The Americans were off to fight in another war against the Japanese, but, within a few years, they would return to Manston—although that is another story.